April 2004

For Bob & Annabel,

Hope you enjoy this
Colorado journey.

Keith Holmshd.

# COLORADO
## Visions of an American Landscape

**T. C. MILLER (attr.)** Hydraulic Mining at Alma, ca. 1881. *Courtesy of the Denver Public Library Western History Department.*

# COLORADO
## *Visions of an American Landscape*

*text by* Kenneth I. Helphand
Ellen Manchester, *Photo Editor*

ROBERTS RINEHART

**ROBERTS RINEHART PUBLISHERS** in cooperation with the
Colorado Chapter of the American Society of Landscape Architects

*Sponsored by the Landscape Architecture Foundation*

*Dedicated to our families:*
*Margot, Sam and Ben*
*Bob and Walker*

Copyright © 1991 by the Colorado Chapter of the American Society of Landscape Architects and the Landscape Architecture Foundation

Published in the United States by **Roberts Rinehart Publishers**, Post Office Box 666, Niwot, Colorado 80544

Published in the United Kingdom and Europe by **Roberts Rinehart Publishers**, 3 Bayview Terrace, Schull, West Cork, Republic of Ireland

*Designed by Polly Christensen*

International Standard Book Number 1-879373-06-8

Library of Congress Catalog Card Number 91-65626

*Printed in the United States of America*

*This book was made possible through the generous donations of funding by the following individuals and organizations:*

Martin J. and Mary Anne O'Fallon Trust
Members of the Colorado Chapter of the American Society of
  Landscape Architects
Landscape Architecture Foundation
Carl A. Norgren Foundation
The Colorado Endowment for the Humanities, Denver, Colorado, a
  grantee of the National Endowment for the Humanities
Denver Water Department
Colorado Horticulture Research Institute
Ruth and Vernon Taylor Foundation
Mrs. Hugh Catherwood
Ms. Elizabeth Schlosser
Mrs. C. L. Hubner
Mr. and Mrs. Don D. Etter
Ms. Jane Silverstein Ries
Phillip E. Flores Associates, Inc.
Design Studios West
Mr. and Mrs. Walter A. Koelbel

# Contents

*Preface* ix
*Acknowledgments* xi
*About the Photographs* xv

## LANDSCAPE THINKING xxi

## PART ONE SPACE, LINES & SECTIONS 1

The Colorado Rectangle 3
Lines 5
Cross Section 9

## CHARACTER 29

Primary Elements 29
Vista 33
Mountains 37
Passes & Gateways 43
Frontier 45

Boom & Bust 49
Ghost Landscapes 53
Sublime & Beautiful 55
Mining Metaphor 64

## EXPLORATION 67

Quests & Questions 67
First View 69
Maps 75

## PART TWO    ROCKS    83

Rushes    85
Diggings    95
Tailings    105

### WATER    113

Patterns and Perceptions    113    Diversion    131
Ditches    119    Doctrine    137

### PLANTS    143

Farm & Field    143
Ranch & Range    159
House & Garden    169

## PART THREE    CONNECTIONS    175

Access    177    Roads    185
Tracks    179    Sights and Sites    195

### SETTLEMENT    215

Cliff Dwelling    215    Colony    232
Plaza    219    Main Street    235
Camp    221    Denver    237
Boom Town/Ghost Town    229    Ski Resort    243

## VISIONS

Powell    249
Parks     253

La Vega   262
Re-       264

*Bibliography*    267
*Photography*     276
*Index*    279
*Index to Photographers*    284

# Preface

Change has aways been apparent on the lands of Colorado. The dramatic natural processes of erosion of the mountains, landslides, floods, drought, so alter the landscape that the stories of those events can be read in the land for generations. Human influences—settlement, rails and roads, agriculture and mining—leave significant marks on the landscape. The state's dramatic vistas and crisp, clear air, and the spareness of the landscape offer the opportunity to see clearly the history of Humans and Nature in Colorado.

In the late 1970s, Colorado's sudden population increases began a cycle of growth and change that would, like other significant periods of development, once again dramatically reshape the landscape.

Landscape architects study how the landscape evolves and endeavor to influence it through the work of the profession. As the members of the Colorado Chapter of the American Society of Landscape Architects saw in the late 1970's, the forces then changing the landscape extended far beyond the profession's influence. The quickness and the significance of the changes convinced them that the nature of Colorado's landscape and the forces causing change should be understood and recorded. CCASLA formed a committee of its members to define the project and to develop support for it. The committee proposed a book of such depth and breadth, that the project was undertaken in steps.

The Colorado Horticulture Research Institute and CCASLA funded early research. With the addition of a planning grant from the Colorado Endowment for the Humanities the committee hired Carolyn Etter as project director, selected the author and photographic researcher, and developed a budget and detailed outline of the book. The Landscape Architecture Foundation provided financial coordination and their non-profit status, along with help and encouragement. Scores of individual members of the CCASLA volunteered time and expertise in fund raising, publicizing, and planning. Chapter presidents, from 1979 through 1991, served on the Book Committee, offered support and inspiration, organized events, and raised money. Kenny Helphand and Ellen Manchester waited patiently until all the money was raised, and then made room in their busy schedules to create this work. Without the help and guidance of these individuals and organizations, and but for the heroic commitment of the members of the Book Committee, this book would not have been possible.

William Wenk

# Acknowledgments

FROM ITS INCEPTION this book has been a collaboration. The book committee of the Colorado Chapter of the American Society of Landscape Architects selected Ellen Manchester and me as author and photographic editor. In this "arranged marriage" we discovered mutual and complementary expertise in photography and landscape and the joys of working together. In the process we learned much from each other and the results of our many discussions are embodied in our photographic choices. This work represents that interaction and we believe it to be the first such landscape and photographic history and interpretation of a state.

The other essential collaborator has been the CCASLA book committee. They conceived the idea for the book, gathered the necessary resources, and assured its progress at every step. I am not a Coloradan, but offer the perspective of a scholar of the American landscape, raised in the East and residing for almost two decades in the Northwest. I knew "landscape", but the committee was indispensable in helping me learn Colorado. Their diverse viewpoints, generous hospitality, good humor, insightful reading, and helpful hints were all valuable and are represented on these pages. Special thanks are due the committee's chairs Caroline Etter and Ruth Falkenberg and also to Bill Wenk who for years shepherded the project along and was a continual catalyst. Rick Rinehart, the publisher, was also a committee member and has been of immense help. Like guiding parents the committee knew exactly when to let go and allow the book grow in its own direction. For all this has a been a labor of love and concern for the Colorado landscape.

Many others were indispensable: those who assisted during travels along the way: Robert Adams, Herb Schall, Felicia and Paul Wilbert, Gene Bressler, and the many financial supporters. I am particularly appreciative of the efforts of the many readers of the work in progress. In the final stages D. Teddy Diggs provided superb, sympathetic and authoritative editorial assistance and Polly Christensen a wonderful classic design for the book

Source materials are silent collaborators in the process of research. First and foremost was direct experience with the Colorado landscape over thousands of miles of travel through all parts of the state, through towns of all sizes, on Main and back streets, on and off road, from peak to valley, and inumerable museums and coffee shops. The landscape itself was not only the object of study, but a source of guidance and pleasure, offering direction, and provoking questions. Second, was the landscape experience of others represented through journals, diaries, autobiography and literature. The struggle was to mold the personal and local to the broader history of the state and its times. The final product is impossible to imagine without an intellectual and personal debt to J. B. Jackson whose writing sets the standard for understanding the American landscape and who many years ago urged me to go west.

Finally my appreciation to my family, Margot, Sam and Ben who have taken pride in this work, accompanied me on my travels and shared my enthusiasms. I see through their eyes as well.

**Kenneth I. Helphand**
*Eugene, Oregon  August 1991*

*The members of the Book Committee of the Colorado Chapter of the American Society of Landscape Architects and the authors would like to gratefully acknowledge the contributions of expertise and support from the following individuals:*

| | |
|---|---|
| Robert Adams | Patricia J. Harrington |
| John Albright | John B. Jackson |
| Nancy Baker | Robert Jackson |
| Lois Brink | Charles G. Jordan |
| Francis W. Brush | Richard D. Lamm |
| William Cronon | Robert Z. Melnick |
| John Dillavou | Terry Minger |
| James B. Dixon | Gerald Patten |
| Carolyn Etter | James C. Pierce |
| Don Etter | John T. Schwartz |
| Janis Falkenberg | Jane Silverstein Ries |
| Phillip Flores | Frederick R. Rinehart |
| Glen Fritts | Guenther Vogt |
| Bonnie Hardwick | William Wenk |

# About the Photographs

COLORADO HAS A RICH photographic tradition ranging from the spectacular nineteenth century landscapes of William Henry Jackson and the quiet, pictorial work of Laura Gilpin to Arthur Rothstein's bitter documents of failed farms in the 1930's and Robert Adams' haunting and provocative suburban and rural landscapes along the Front Range. In this book the photographic medium is explored as a record of the development of our culture's attitudes toward landscape for we can no longer separate landscape images from landscape experience. This book places this visual history in the context of the development of landscape values in Colorado.

The history of photography roughly parallels the history of Anglo-european settlement of Colorado. Records trace the earliest photographic efforts to document the Rocky Mountains and western territories to J. Wesley Jones's expedition of 1851, only twelve years after the official invention of photography in France. Although no actual plates are known to exist, reproductions and accounts in popular journals of the day refer to the more than 1,500 daguerreotypes he made in the region. The unwieldy daguerreotype process was also used by the artist Solomon N. Carvalho when he accompanied Colonel John Fremont on his final expedition to the western territories in 1853.

Although the photograph was often promoted as the true document, free of artistic embellishments and pictorial conventions, it also vied for recognition in art circles as creative expression equal to that of any painting of the day. Photographers and the pictures they made were inextricably bound to the cultural beliefs, artistic fashions and political climate of the period. The spirit of Manifest Destiny and its aggressive western expansionism; pictorial conventions of the sublime, idyllic landscape learned in European academies; and the myth of the romantic hero/ mountainman in the wilderness were all strong influences on the resulting depiction of the West as a place of exotic beauty, of wildness to be tamed and controlled, and of landscapes of monumental proportions and of immense resource development potential.

Carvalho's account of photographing from a Colorado mountain top reveals much about the attitudes these artists brought with them from the East Coast, and surprisingly little about the ordeal of making daguerreotypes standing in snow up to his waist with temperatures often plunging to 20 - 30 below zero.

**HORACE S. POLEY** Taylor Park Placer Mining, ca. 1895. *Courtesy of the Denver Public Library Western History Department.* In this image the landscape becomes a metaphor for the settlement of the West. The dramatic mountain backdrop sets the stage for the mining party perched at the edge of the "frontier," looking into the future.

## About the Photographs

*"... after three hours' hard toil we reached the summit and beheld a panorama of unspeakable sublimity spread out before us; continuous chains of mountains reared their snowy peaks far away in the distance, while the Grand River [now the Colorado] plunging along in awful sublimity through its rocky bed, was seen for the first time. Above us the cerulean heaven, without a single cloud to mar its beauty, was sublime in its calmness. Plunged up to my middle in snow, I made a panorama of the continuous ranges of mountains around us."*

**Solomon S. Carvalho**  Incidents of Travel and Adventure in the Far West; With Col Fremont's Last Expedition Across the Rocky Mountains. 1857

After the Civil War the country was anxious for a new national identity, and systematically sought it in the monumental landscapes beyond the Mississippi and Missouri Rivers. New photographic processes had been thoroughly tested in the extensive documentation of the war, and many of the photographers took their skills to the West. Public interest in photographs of the exotic, grand and sometimes austere landscape of Colorado grew rapidly as the images were widely distributed throughout the East Coast in the form of stereo views or reproductions in popular journals of the day. Stereo views were small photographs that when viewed through a special holder produced a three-dimensional image. Sold as sets, one could subscribe to a series and be an armchair traveler through the "Rocky Mountain Scenery" or Garden of the Gods and other natural wonders. The public enthusiastically embraced the absolute veracity of these photographs even in the form of a magazine reproduction—of an artist's interpretation—of the original photograph. To a country ripped apart by war, with little appreciated cultural history and few architectural treasures, photographs of the West began to describe a nation of awesome strength, sublime beauty, boundless frontiers and unlimited natural resources. Colorado literally became the gateway to the Rockies and, through photographs, soon became the symbol for everything the West represented in the eyes of Easterners.

The photographer whose work has probably had the most influence on how we define the Colorado landscape is William Henry Jackson who worked in Colorado from 1873 to 1897. Jackson, who was trained as a painter and once worked side by side with the painter Thomas Moran, relied heavily on romantic pictorial conventions of the period to create his dramatic images of the new West. The legacy of Jackson's work in Colorado can be seen today in postcards, travel brochures, Sierra Club calendars, climbing magazines and advertisements promoting the Colorado experience. In fact, many of these images, such as Maroon Bells, Garden of the Gods, and Royal Gorge have been made from the exact location of Jackson's original photographs more than one hundred years earlier. His photographs literally direct how we experience that landscape today. The view, and the experience, become validated and commodified as picture spots. Jackson also photographed the marvels of the new industrial era, particularly mining and the railroad. Here again nature became commodified, seen as a resource to be exploited and developed through commerce and tourism.

Jackson's influence can also be seen in the work of his former assistant Louis Charles McClure whose career spanned more than sixty years in Colorado. McClure followed closely in Jackson's tradition photographing rail lines, scenic views, cityscapes and agricultural scenes to promote tourism and development in the state. McClure's simple, direct style and unusual vision made common places such as mining towns, river valleys, ranches and rail lines places of exceptional beauty. His photographs were widely distributed and published, and contributed significantly to the portrait of Colorado at the turn of the century.

In the late 1930's Farm Security Administration photographers documented the effects of the Depression, the dust bowl and the ravages of eighty years of mining on Colorado's land. In sharp contrast to Jackson's romantic landscape views and images which exalted the mining and railroad industries, these images described a region seriously out of balance in its relationship to the land. Interestingly, Marion Post Wolcott was sent to Colorado by the FSA in the early 1940's to make photographs that would show America's agricultural bounty and the virtues of small-town America.

For the past thirty years the photographs of Robert Adams have challenged the romantic myths and pictorial conventions of the nineteenth century. A native of Colorado, Adams has doggedly pursued a quiet yet provocative portrait of Colorado's Front Range and eastern grasslands. His photographs of suburban and rural Colorado question many of our preconceived notions about the romance of the new west and the ongoing myth of the frontier.

The contrast between Jackson's highly stylized images and the hard social edge of the FSA work is indicative of the broad range of photographic activity throughout the state's history. Both natives and visitors alike have contributed to this remarkable visual record which includes family snapshots and anonymous highway department documents as well as the work of such nationally recognized artists as Ansel Adams and Jerry Uelsmann. This book seeks to place the work of these photographers in an historical and cultural context in order to better understand how their images reflect and shape our attitudes about the Colorado landscape.

Several principles guided the selection of these photographs. First, landscape photographs were defined by images that represented the relationship of people to place. Second, each photograph tells more than one story, speaks to several levels of landscape thinking, and often represents similar issues in another period, i.e. a nineteenth century mining photograph might address contemporary environmental concerns. Third, the selection attempts to provide a sense of the history of photography, specifically as it relates to landscape and cultural issues. Various styles of photography from every decade since 1859 are represented by a diverse group of more than sixty photographers. To reflect an acurate history of the medium, a wide range of types of photographs was selected, ranging from the snapshot to government documents and fine art images. Of particular interest were photographs that were previously unpublished or little-known. Fourth, in an attempt to avoid "Denver-centrism" or "mountain-centrism," images were researched and collected from all parts of the state, to represent the rich variety of landscape spaces within the region. Fifth, the selection of photographs by and about the cultural diversity of Colorado's landscape history was an ongoing consideration. Photographs by and about minorities and women figure significantly into this history. Sixth, the photographs are to be seen for their distinct narrative quality in their own right, not necessarily as illustrations of the text. In addition, the text does not always describe specific images. Both are designed to complement each other and to stimulate further thinking.

Many people and institutions have been instrumental in the collection of these photographs.

We visited the archives of numerous small historical societies, public libraries, and museums in the four corners of the state and throughout the Rocky Mountains, Front Range and Western slope. Although constraints of staff and facilities limited the number of images from these smaller organizations, their importance can not be emphasized enough in terms of influencing the final selection of photographs for this book. Close to half of the historical photographs have been drawn from the impressive collections at the Denver Public Library's Western History Department and the Colorado Historical Society's Photographic Collection. I know of no other state that has such extraordinary photographic resources that are so well managed and so accessible to the public.

Eric Paddock and the staff of the Colorado Historical Society generously assisted our research with enthusiasm and knowledge. He would steer us to unusual, little-known work, often opening collections that had yet to be catalogued. No family album or highway department file was considered unworthy of our perusal. The scope, content, quality and accessibility of this collection is extraordinary by any standards.

Augie Mastroguiseppe and Kathey Swann of the Denver Public Library's Western History Department brought cart after cart of photographs for us to research, and spent endless hours on the phone tracking down obscure images recalled from five-year old notes. In an age when its hard to imagine information retrieval without elaborate computer systems, the staff of DPL's Western History Department always found the images we were looking for, and often offered alternatives for our consideration. The collection is superlative in its range of original photographic material.

We are indebted to photographer Robert Adams who gave his support and encouragement in the early stages of discussion with the committee. His precise, critical thinking on landscape issues has served as a valuable guide in the editing of these photographs.

Finally, very personal and heartfelt thanks go to Caroline Hinkley for providing bed, board, strong coffee and late night discussions about landscape and politics. We have shared much of the Colorado landscape together for more than 13 years.

This book has been a integral part of my family life since my husband, Robert Dawson and I photographed toxic waste sites throughout the American West on our honeymoon in 1983. That summer, and for several to follow, we journeyed throughout Colorado visiting archives and literally making thousands of negatives from mountain top to prairie. Much of my appreciation for the richness of the Colorado landscape has been learned from looking through the viewfinder of his camera. And to my son Walker, whose arrival actually made the completion of this project possible—I hope he will enjoy this book.

**Ellen Manchester**
*San Francisco*
*August 1991*

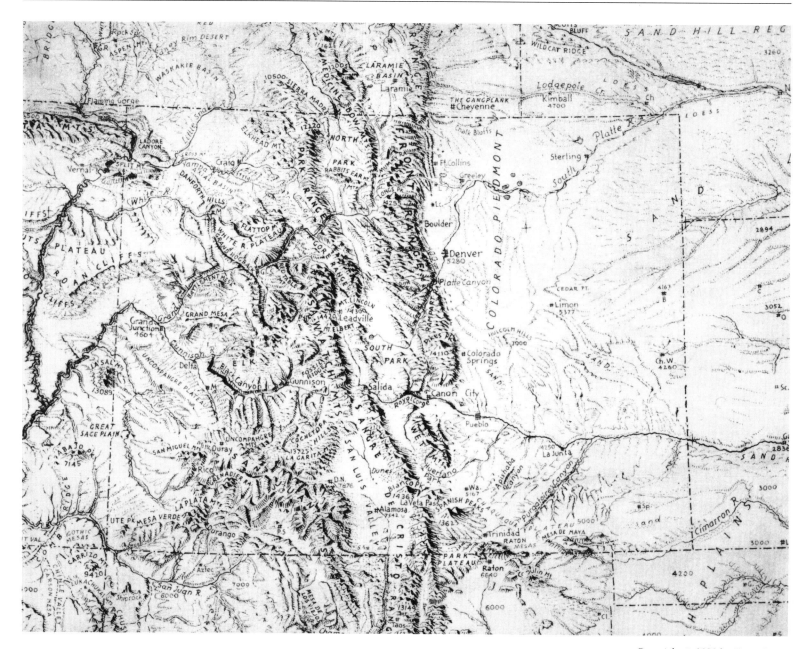

Copyright © 1939 by Erwin Raisz.

# INTRODUCTION: LANDSCAPE THINKING

This is a work of landscape history, landscape understanding, and landscape appreciation guided by the following principles.

**The interaction of people and place is embodied in the creation of landscape.** Landscape is not scenery, although it includes the scenic, nor is it the natural world without the human presence. The landscape is a creation, the record and repository of the discourse between people and the physical environment.

The word *landscape* has a dual origin and multiple meanings. The English word derives from *landskip* and *landschaft*. *Landskip* referred originally to paintings that depicted the domesticated countryside as a scene, with pictorial conventions derived from theater and painting. *Landschaft*, on the other hand, was a collective term for the whole of village, gardens, fields, and woodlot. It was the world of the rural resident. Modern usage emphasizes the landskip meaning, the visual and scenic, but modern times demand a return to other origins, to landscape as a social concept, the bond of people and place.

The sources for landscape study are many, but the primary source material is the landscape itself, for in its form is the record of its history. In J. B. Jackson's dictum, "Landscape is history made visible."

Why does the landscape look like it does? How did it assume that shape and form? What is its identity and structure? What did it mean? What does it mean now? There is a continuity to landscape history, undoubtedly due to basic responses to natural forces. People in each era inherit a landscape, which they then pass on, yet certain formative periods and developments are critical in landscape history. They set the parameters of activity, the stage and framework for subsequent developments and landscape events. For example, the choice of a settlement site or route is generally reinforced by later generations with little wonder about the crucial initial selection.

History is personal. Human biographies and landscape biographies describe a past. A portrait of a place, like that of a person, can capture a personality. Each individual has his or her own intimate

**WILLIAM HENRY JACKSON** Moraines on Clear Creek, Valley of the Arkansas, 1873. *Courtesy of the United States Geological Survey*, Denver. Jackson became Colorado's pre-eminent photographer of the 19th Century. While much of his government work in Colorado in the 1870's recorded the spectacular scenery of the Rocky Mountains, many of his photographs also documented sites such as this which had the potential for future water or mineral development.

**MARK KLETT and JOANN VERBURG for the Rephotographic Survey Project**  Clear Creek Reservoir, 1977. *Courtesy of the Rephotographic Survey Project.* The Rephotographic Survey Project located more than 200 sites of 19th-century government survey photographs of the American West and remade the images from the exact camera position, at the same time of day and year, approximately 100 years later. The project documented major geological and vegetational changes, as well as the effects of human settlement on the land over the 100-year period.

story, which is inseparable from the specifics of place and time. Like politics, all landscapes are local. In autobiographies, journals, and first-person records of the Colorado landscape experience, we come closer to directly experiencing a place and situation. Most important, these accounts are a reminder of the intimate connections we all have to places, connections that are often difficult to articulate, for they are visceral and emotional.

Colorado adults know well the measures of landscape change, for memory acts as a gauge and guidepost. Many recall moving from a city to a suburban fringe, an outpost of community bordered by foothills, rangeland, or desert, only to later see favored and secret places of play and discovery subdivided, paved, and culverted. The busy downtown of stores and offices was abandoned. Urban walking gave way to urban fear. One can generalize from the particular; through autobiographies, the history of the landscape is revealed. Despite those who proclaim "future shock," this rapid pace of change had been true for all of Colorado's century-and-a-half modern history. It was certainly equally true for the native inhabitants, miners, farmers, or ranchers as it is for today's suburbanites. The Colorado landscape is largely the product of the modern world, its technology and imperatives, which recognizes no boundary of state or nation. Yet, these elements take on a cast and character in the specifics of Colorado places.

Landscapes carry legacies and lessons. Legacies are the inherited evidences of tradition and mores. Not all are to be respected or cherished, but they demand to be understood. Lessons may instruct us, remind us of original intentions, help us avoid mistakes, clarify our ideas, or act as an inspiration. Oftentimes, good ideas are lost, only to resurface or reappear in a period more receptive to their potential.

**Landscapes are understood at a variety of spatial and temporal scales.** The experience of the Colorado landscape is evidence of great continental geological processes, miniscule sites, and ephemeral events. The landscape can be approached on several levels, which form a continuum. At one extreme it is possible to discern patterns that are universal or archetypal, those that may be common to all cultures as they settle and transform the land. Next is the national, the development of national character and culture and its manifestations in the landscape. Then comes the regional level, areas bounded by a common heritage and environment, and the local, the minutia of places and their matrices of meaning. Finally, there is the personal landscape—at the other extreme, the idiosyncratic, the wonderful peculiarities of individual personality.

A complete understanding of any landscape necessitates spanning these levels, for the boundaries are not sharp and there is a resonance among them. However, a focus is also instructive. This book is about Colorado, a state, which is part of a larger national system and is itself subdivided into regions. The emphasis in this work is on understanding the history of the Colorado landscape as it embodies characteristics of American culture, values, and life and how they are given distinctive form and character in Colorado.

**Landscapes are texts to be "read."** However, unlike the authorship of most books, the authorship of landscapes is collective. Landscapes speak in languages and dialects. Landscape literacy necessitates an understanding of landscape vocabulary, grammar, syntax, and modes of expression. Like texts, landscapes have a composition, patterns of expression, and they impart messages. Different cultures with distinct values and behaviors have created the Colorado landscape. Their values, beliefs, and wishes are visible on the land. This book is one reading of the Colorado landscape. It does not exclude others. The intention is to encourage multiple landscape readings, for reading is an interactive process between authors and readers.

Reading the landscape involves dilemmas. There is the question of what to read. Does one look at soils, vegetation, houses, signs, colors, materials, plants, patterns, or things that are big? Not everything is faithfully recorded. Landscapes record certain characteristics of a people better than others. Some messages are made, then rapidly lost, for the materials of landscape decay, grow, and transform. Often, in our analyses, we are prisoners of what persists, although the ephemeral may have greater significance than the permanent. Current concerns and questions may not be those of the past. We may focus on what was insignificant for those in the past and miss matters of great import. Therefore, reading the landscape directly is best augmented and accompanied by accounts of the time, to help us understand the connections between intentions and result.

**Landscape criticism is essential.** Although Colorado suffers from profound environmental problems, this work is not a polemic but a historical study. The book is guided by principles of honesty, tolerance, pluralism, and a respect for places and for human interaction with those places. The foundation of a responsible criticism is a recognition of existing conditions and how they evolved. Only then is informed analysis, evaluation, and decision making possible. It is in that spirit that this work is offered, as one modest step toward the creation of an informed landscape citizenry.

The observations of two critics have helped guide my approach. Northrup Frye, the Canadian literary scholar, divides the critical process into two related yet distinct parts. Frye notes: "The fundamental critical act . . . is the act of recognition, seeing what is there, as distinct from merely a Narcissus mirror of our own experience and social and moral prejudice. Recognition includes a good many things, including commentary and interpretation." Harold Clurman, a theater and literary critic, notes: "The first step in criticism is to see, to feel, to absorb the work presented. . . . The second step in criticism is judgement or evaluation." This book is directed at helping us see and recognize the Colorado landscape.

**Design is the act and art of giving form to ideas.** Landscapes are influenced by design professionals and policymakers, but landscapes are created mostly by the gradual accretion of countless human actions. We are all designers, and all designs are worthy of investigation and analysis. Design is a process and a product, a verb and a noun. The processes of design and planning shape the landscape. They are part of the mechanisms of change, the willful manipulation of cultural processes. Designs are created, but to design is also to desire. Designs are actuality, but they are also visions and aspirations.

Different cultures with distinct landscape values and behaviors have created the Colorado landscape. Responses to the land have varied: husbandry, worship, civilizing, speculation, despoliation, preservation, stewardship, recreation, settling. The interactions of people and place are embodied in the creation of a landscape.

# PART ONE

*Space, Lines, & Sections*

*Character*

*Exploration*

# SPACE, LINES, & SECTIONS

## *"This scenery satisfies my soul."*

THE LANDSCAPE IS THE BASE, the foundation, for all human activity. We reside on and with the land. It is also a frame, the demarcated territory that organizes, structures, and defines our lives. The history of Colorado is landscape history. In this land of 14,000-foot peaks, deep canyons, immense spaces, hidden valleys, and grand mesas, getting to places, servicing places, communicating between places, and making a living in places have all left their mark on the land. For explorers and surveyors, the land was a mystery, presenting a challenge of physical and intellectual dimensions, as both real and psychological terrains had to be traversed and confronted. The sheer fact of crossing and moving through the land was a drama of trails, passes, hardship, perseverance, imagination, and engineering. The native inhabitants knew the land well: its limits, possibilities, character, and glory. The settlers of the past century and a half had to relearn, and are still learning, these landscape lessons.

Landscapes have geometries of points, lines, and spaces. The four sides of the Colorado rectangle are a work of political artifice, yet they define more than political boundaries. The arbitrary straight lines, if applied loosely, cut through natural and cultural regions, landscape spaces shared by Colorado's neighboring states in the American West. The eastern third of the state merges and overlaps with the plains and prairies of Kansas and Nebraska; the southern section joins with the southwestern desert and the Hispanic-Indian-Anglo culture of New Mexico; the western Colorado plateau is a geographic domain shared with Utah and Arizona; and at the northern border, the landscape resembles the dry, harsher ranchlands of Wyoming.

Colorado culture is a product of this overlap, yet there is a central core. Geographically,

## THE COLORADO RECTANGLE

**Photographer Unidentified**
Sterling, View From Southeast.
*Courtesy of the Denver Public Library Western History Department.*

4

topographically, and spiritually, the Rockies dominate the landscape of the state. Demographically, economically, socially, and culturally, the companion communities of the Front Range, particularly Denver, are dominant. From this nucleus of power and population it is often easy to take a "Denvercentric" position and forget the hinterlands, yet a reading of any map shows the futility, even the danger, of such an action. Despite the political boundaries that influence our lives, the natural landscape base is a continuous system. Colorado's magnificent mountains are only portions of great continental chains. Here lie the headwaters of four great river systems: Platte, Arkansas, Rio Grande, and Colorado. National transportation systems of trails, rails, roads, and airways converge on the state, as do the contemporary environments of mass media. People, goods, and ideas have origins in specific places, but they also have lives of their own, which often ignore physical imperatives and political boundaries. Colorado is linked to the surrounding states and the nation, and in many ways the cultures and landscapes of these communities merge and overlap. Yet, despite this fact, one can also describe and characterize a distinguishable Colorado entity.

**ANSEL ADAMS** Maroon Bells, Near Aspen, 1951. *Copyright © 1991 by the Trustees of The Ansel Adams Publishing Rights Trust and The Friends of Photography.*

## LINES

The entries to Colorado communities, the transition points from the highway to town, are marked with signposts, with obligatory listings of lodges, service organizations, churches, and oftentimes some claim to local fame, such as "Manassa—Birthplace of Jack Dempsey." The official state sign enumerates the town's elevation, not the population, as is often customary elsewhere. In Colorado, height above sea level is the key indicator—how high can reveal much more than how many. The range is grand—from Denver, the Mile High city, to mountain passes at two miles, to peaks at almost three vertical miles. The mile-high benchmark is engraved on the steps of the state capitol. The state's average elevation is 6,800 feet. With the vertical changes in elevation, a set of parallel events occur as you move up or down Colorado's slopes and valleys. As you ascend, the land gets rockier, the air thins, the temperature decreases about three degrees per 1,000 feet of elevation, precipitation and wind increase, the growing season shortens, habitat and the life zones change. Rising 1,000 feet is the rough equivalent of traveling north six hundred miles. This all happens with a change in elevation, and plants and animals adapt to these different

circumstances. People similarly try to adapt, sometimes respecting the imperatives of their environmental circumstance and ameliorating their effects, but often simply attempting to deny these effects.

    The timberline is the upper limit of continuous tree growth. (The tree line is the limit of any tree growth.) One can look at the Colorado landscape in timberline terms. There are two timberlines, an upper and a lower. The timberlines are visible from the communities of the Front Range and from valley floors. Virtually all woods and forests exist in a horizontal band between 5,600 and 11,000 feet, a green stripe one moves into and beyond. Beginning the descent at higher elevations and passing through the upper timberline, one moves down mountainsides. The rock face of the land is paramount, the ruggedness visceral, and the sky close. This is an alpine tundra, a fragile, fractured rock garden. Moving from this dramatic but unprotected landscape, one passes through the membrane of Krummholz forest (German for "crooked wood"), into the protection of the woods, the cover of green, and the security of tree trunks. For a vertical mile the passage is through successive zones of coniferous forest. A subalpine zone over 9,000 feet is dominated by Engelmann spruce and subalpine fir and the seasonal transition to glacial lakes. Then forests of ponderosa pine found on drier slopes and Douglas fir in wetter areas take over, with broken stands of aspen and lodgepole pine. In the lower, drier mountains of the West Slope, piñon pine predominates. The eventual downward passage through the forest leads to the second timberline. Located in the foothills, this is a transitional zone of great natural diversity, as are all ecotones, the territories of boundary and overlap.

    The two timberlines mark three zones. The upper zone is the peaks, the spiritual heart of the state, the Rockies that preside over all, the points of orientation and reference. These are seen but only occasionally visited. In spring and fall this zone is marked in white. The middle zone is where much of the state's wealth has come from, first in minerals and now in tourism and recreation. The lower zone is where people dwell. It is the agricultural state, the ranching state, the urban state.

    Additional "timberlines," boundaries between wooded and open landscapes, are apparent as well. There is the pronounced riparian edge of cottonwoods and willows along the plains river valleys, the Platte and the Arkansas. There is the green line surrounding settlements,

**JOAN MYERS** Iron Springs Stage Station, 1983. *Courtesy of the artist.* This stage station was on the Santa Fe Trail. The post stubs are all that remain of the rectangular corral that enclosed the station's livestock.

**ROBERT ADAMS** Genoa, 1970. *Courtesy of the artist.* "Prairie buildings—spare, white, and isolated—are emblems of our hope and its vulnerability. Even architecture in town finds its reference point at the ends of streets, at the horizon; we sense there an expanse so empty that it can almost seem to spin. Mystery in this landscape is a certainty, an eloquent one. There is everywhere silence—a silence in thunder, in wind, in the call of doves, even a silence in the closing of a pickup door. If you are crossing the plains, leave the interstate and find a back road on which to walk; listen." Robert Adams, from *Prairie, Photographs by Robert Adams.*

a line dramatically apparent in the eastern plains. Finally, at the most domestic and personal level, there is the grove, marking and surrounding a dwelling in an open landscape.

Timberlines are natural lines in the landscape, but the land is marked by a great series of lines that are the product of human artifice. They traverse the state. The national grid begun in the Midwest marches across the continent. Its great section lines cut across the plains and prairies, with roads at mile-wide intervals. All of its lines, roads, fences, ditches, furrows, and wires run toward the cardinal directions. Occasionally the straight linearity is broken as watercourses or sharp breaks in topography intercede. Towns emerge as finer-grain grids and city blocks assume the same shape as fields. The straight lines continue up to the Front Range. Long strands enter the Front Range cities, where urban blocks spread through the landscape. The squares of section lines now contain the recent variants of the suburban curves and lollipop cul-de-sacs. In the eastern section, only the major rivers, the Platte and the Arkansas, modify the course of the lines of movement, the roads and railroads. But the mountains are more formidable and overpowering, forcing roads and tracks to wind, curve, zigzag, and switch back. Finally, these routes are compelled to tunnel mountains. To get up and over, they had to go through. On the map the lines look like piles of string, slithering and curving, straightening again only in the long valleys, open parks, and flat mesas. Other lines, the gentler lines of property boundaries, fences, poles, wires, and fields, can often manage the mountainous slopes, running straight across contours and ignoring the imperatives of terrain. Oftentimes, lines run together in crisscrossing braids of river channel, road, railroad, ditches, telephone wires, and fences. The lines make the invisible visible. They mark territory, structure communications, record transactions. Like string, they bind the community together.

Landscapes can be described, studied, understood, and imagined in many ways. One of the most powerful methods is as a cross section. The cross section simulates traversing through space. Sections can be taken horizontally across the land, east to west, south to north, or vertically, from bottomland to peak. The method is a vehicle to describe and evoke the space and spirit of Colorado's distinctive landscapes: prairie, plains, Front Range, Rockies, parks,

**CROSS SECTION**

**L. D. REGNIER**   On the Arkansas, ca. 1900. *Courtesy of the Denver Public Library Western History Department.*

plateau, valleys, mesas. Like a tricolor flag, the state is divided into three landscape bands: the eastern plains, the mountains, and the Western Slope.

Let us envision a cross section, taking a transverse section east to west, following the sun and the course of Anglo-American settlement. In eastern Colorado, soil and sky dominate, along with the vegetation that seasonally cloaks the earth. There are few trees or rocks. The plains are a dry steppe landscape, a transitional belt between the more humid climates of the east and the desert. The colors are browns, the tans of tilled earth, and dry greens, the colors of sage and the short grasses of the native prairie and rangeland, or the deeper greens of crops. The plains are expansive; their appeal is in the drama of vast spaces and the closeness of the ground, the breadth of the sky. Summer shade is a precious commodity found along watercourses, ditches, farms, shelterbelts, and occasionally along roads.

The landscape often looks empty, seemingly devoid of the human presence, but in fact it is all plowed, harrowed, disced, and tilled, almost every square foot. Beneath the soil surface lies an underground landscape of hidden tunnels and burrows of prairie dog towns and dens for snakes, lizards, and small mammals. These sit atop sedimentary rock layers a mile to two miles thick of sandstones, shales, and limestones. It is a much modified landscape. This was the home of the Arapahos and the Cheyennes, who were removed and exiled in the nineteenth century. Here the bison roamed, only to be replaced, along with the prairie grasses they fed upon, by domesticated species of plants and animals.

The measure or rhythm of the land, its pace, is visible in the space between—in the distance between telephone poles, power lines, ranches, farms, windmills, water towers, towns, and creeks. The landscape may be tightly woven in a dense fabric or open and expansive in a loose weave. On the plains, the fabric is loose, and the density, a measure of relative distance, is low. The points of concentration, the aggregations, are often verticals and read like a bar graph of impact, in water towers, grain elevators, sugar beet mills, or the green swatch of a town on the prairie. The high points are the signs of human presence, even the modest windmills that allow artificial water holes for cattle.

The flat horizon line of the plains is punctuated by the skyline of rural communities: cooperative grain elevators; water towers; an occasional radio, microwave, or telephone tower. Towns often have two vertical landmarks, subregional points of orientation. First are

**ROBERT ADAMS**
Newly Occupied Tract Houses, Colorado Springs, 1974. *Courtesy of the artist.* The photographs and writings of Robert Adams have had a profound influence on an entire generation of contemporary landscape photographers. Years before revisionist history was in vogue, Adams challenged the values which the Anglo-European culture brought to, and imposed upon, the western landscape. His quiet and provocative images have dramatically shaped the way we see the "new West." Adams resides in Longmont, Colorado.

the grain elevators, looking like great pillars or pylons supporting the land between them. Along the rail lines of the Burlington Northern Railroad and the Union Pacific they are visible one town to the next. They have an Egyptian grandeur as solitary monuments. Second are the water towers, metal mounts oftentimes surrounded by a park at the base. These structures poke through the soft edge of trees that mark the town. Within the elms and cottonwoods, the town is found.

The flatness is deceptive. There are different plains topographies. They are not plain in any sense; there is great variability. The land undulates in rolling, wavelike hills; there are also bumps, the flatness of billiard tables, and ripples and depressions. The land is falling ten to nineteen feet per mile in a great tilt from the continental divide to the Mississippi. There are many meanings of *plain*. Defined as a level treeless region, the word also means uncomplicated, direct, ordinary, candid, straightforward, pure, and clear. All of these personal characteristics have at times been attributed to the plains landscape and its inhabitants.

Two great rivers cross eastern Colorado and frame the northern and southern plains. In the dry summer, the South Platte and the Arkansas are braided, flat, shallow streams, muddy, with green places in the landscape, riparian ribbons now controlled and much modified by upstream dams. In spring, replenished by melting snows, they become broad surging flows. Looking across the Platte Valley, you see a landscape of layered bands: the sky, the distant grazing land, the soft green of cottonwoods and willows along the unseen river, the regularity of irrigated fields. Harold Hamil, in his high-plains memoir of ranch life in Logan County, wrote: "Because vistas were long and unobstructed from most any point in the valley or along its edges, the South Platte was an overwhelming presence. One lifted his eyes and there, in a panorama measured in miles, was some reminder of the river and its containing valley. It's generally that way in the Great Plains; there seems little to see, yet one never looks without seeing much."

East of Greeley, grazing lands give way to irrigated fields, often in sharp boundary and clear contrast. They are a powerful mark of the human presence and will. Water is transported everywhere—even across dry terrain—en route to its ultimate destination. It is channeled,

**DAVID PLOWDEN**
Leadville, 1973. *Courtesy of the artist,* © *1974.*

funneled, pumped, siphoned, and sprayed. Agricultural land clings naturally to the river corridors, but it also sprouts in great circles, the product of pivot irrigation systems, mining water pumped deep from the Ogallala aquifer. Moving east to west, the grain of the landscape becomes finer, its density of elements and events increasing. Weld County, one of the nation's most productive agricultural areas, is a complex grid of corn, sugar beets, and intermittent feed lots, all made possible by transported and diverted mountain water nourishing the rich soil. Across the plains, grain elevators, strung together by the railroad, appear on the horizon—to the west they are silhouetted against the mountains. The slow emergence of the mountains on the horizon is accompanied by a similar increase in trees.

Much of the eastern and central portions of the state is rangeland, populated by few residents and thousands of animals. Cattle are a natural part of the landscape, and they, much like the trees and the settlements, occur in low population densities. Ranches are found buried in the land, embedded deep within their domains. Portions of the range remain open, allowing the animals free passage, but typically the land is defined to contain the creatures. Cattle guards crossing the roads are subtle reminders of changes in land uses. Fences wrapped around grazing lands have an unanticipated side effect as they emphasize the road corridor and as the ungrazed roadside is transformed into a floral wonder.

At the centerline of the state between Denver and Colorado Springs, the Palmer or Arkansas divide (elevation 7,300 feet) separates the watersheds of the Platte and the Arkansas. On crossing the Palmer divide in 1820, John Bell of the Long Expedition noted, "Naturalists find new inhabitants, the botanist is at a loss which new plant he will first take in hand—the geologist grand subjects for speculation—the geographer and topographer all have subjects for observation." Both river corridors were pathways west. Settlers followed the Platte along the Oregon Trail, and the Arkansas, where farmworkers now tend melons and onions, served as part of the Santa Fe Trail. The southern territory of the Arkansas Valley is drier and lacks the wealth and prosperity of the Platte Valley. The flow of the Arkansas is less than that of the Platte, and at least east of La Junta, the river does not have the same impact on its hinterland. The area is drier. The Prowers County Fair in Lamar is the Sand and Sage Festival.

**L. C. MCCLURE**  The Winding Platte in South Park, ca. 1910. *Courtesy of the Denver Public Library Western History Department.* Louis Charles McClure began his photographic career in 1883 as an assistant to William Henry Jackson in Denver. In 1899 he set up his own business focusing almost exclusively on Colorado views. He built an impressive and comprehensive archive of cityscapes, scenic views along the railroad lines and agricultural scenes all of which were used to promote and encourge tourism and settlement in the state. McClure was not only a highly skilled craftsman, but an artist with an extraordinary vision and an eye for the unusual and provocative.

**MARION POST WOLCOTT**   Yampa River Valley Ranch, September 1941. *Courtesy of the Library of Congress, Farm Security Administration Collection.* From 1935 to 1942 the Farm Security Administration employed some of the county's finest photographers to document rural America. Guided by Roy Stryker, a Colorado native from Montrose, the FSA became the largest and most comprehensive photographic document in history, producing more than 270,000 negatives which are now housed at the Library of Congress.

A line can be drawn across the state between Pueblo and Colorado Springs. U.S. Highway 50 may mark this divide most clearly. The line bisects the state, dividing north from south. Unlike the Mason-Dixon line, it is nameless, but it indicates an equally significant cultural division, demarcating what historians call the borderlands, the northern boundary zone of Spanish influence and Hispanic culture. South of the line, there is a strong sense that a border, a cultural isobar, has been crossed. One notices an increased frequency of Spanish place-names, pickup trucks, sheep, mobile homes, and chiles; more Spanish is heard, and adobe makes its appearance as a building material. Some signs are subtle. The southern towns are much more "open," their yards miniature fields and the landscape more workaday. Residential landscapes start to eschew lawns; stone "mulch," pebbles, and rocks begin to take over. It is a modest but telling indicator. A walk through La Junta Park, a WPA project, reminds one of a Mexican park in its use of stone walls and walkways. The Arkansas Valley is Hispanic, whereas the Platte is northern European, Germanic in its order and crops.

The southern and especially the southwestern regions of the state are oriented to the south, just as the plains look to the east and the state as a whole looks to the west. Colorado represents the northernmost extent of Mexico; the long arm or vestige of history continues centuries after political changes have occurred, for this domain was under Mexican sovereignty until 1848. The residents of the Hispanic and Indian villages of the region represent the state's earliest inhabitants. They are no longer dominant but are recessive; yet under the right conditions, they are capable of exerting tremendous influence.

The Front Range is a continental ecotone, where plains and mountains meet. The mountains stand to one side, the plains the other, but the sides are unequal in impact. For those who trekked across the plains, in wagons or even as truckers or tourists, the plains resided as powerful memory. The plains are horizontal—one needs to pass through them or to rise above them to fathom their full extent. Magnetically, the plains are the negative pole; it is the mountains that beckon. The westward mountains loom, they are visible, their presence is felt; they provide the overview and act as the state's facade. The mountains are the controller, the generator, even manipulating the weather: rain, snow, thunderstorms, hail, smog. Their heights dam the eastward-moving clouds, which drop up to forty inches

**Photographer Unidentified**
Black Canyon of the Gunnison, 1910. *Courtesy of the Colorado Historical Society, State Highway Department Collection.*

**KENNETH HELPHAND**  Great Sand Dunes National Monument, 1989. *Courtesy of the author.*

of precipitation, much of it as yards of snow, on the mountainsides, in turn creating along the Front Range a rain shadow that receives only thirteen to sixteen inches per year.

It is no wonder that people climb the mountain slopes and assault the faces. As if to control or share in the mountains' energies, people have built roads and rails to the mountaintops. "Pike's Peak or Bust" was an exhortation to get to the goldfields, but for an engineering mind, a road or rail to the top was more of a challenge. It is still a scary ride. What compelled such construction? It goes beyond the simple path. The symbol loomed as a challenge to be mastered. The view from the top is of the mountains, but most people choose to look toward the foothills, to Colorado Springs far below and the plains beyond. They measure the distance they have come. The long roads are vague at the horizon, reaching the vanishing point of perspective. "Can't you see Kansas?," I heard someone say. You can't, but you can see over one hundred miles. The ride up has the feeling of an ascension; not surprisingly, a monument to America's Olympic-medal winners has been erected at the summit. The peak is also crowned with a railroad station, a viewpoint, a giftshop, and a parking lot.

The towns of plain and piedmont are compact, bounded by green and set within agricultural or range land. The towns of the Front Range sprawl in tentacular fashion. The metropolitan communities view expansion up and down the range, and to the east especially, as entering into "empty" landscapes. But there are no empty landscapes, only an emptiness of awareness. The attitude to the landscape is, and has been, profligate.

Along the Front Range it feels as if the mountain valleys drain into the plains, which they do. Strands of vegetation come down watercourses; streets and houses finger up the same valleys, following channels and traversing ridges as the lower timberline is lost in a line of domesticated plantings. Settlements are near the mountains, and the foothills timberline has been penetrated by the suburban fringe. Subsumed within city territory, the timberline now often demarcates the urbanized Front Range from the "Mountains." The transitional belt between mountains and plains, forest and prairie, is marked by hogbacks, buttes, and great rock fins. Once an exclusive exurban domain, this urban/rural fringe now contains both ranches and subdivisions that end at fields; farmers, ranchers, and commuters overlap. This

**L. C. MCCLURE**   Fruit Ranch of W. B. Cross, Grand (now Colorado) River Valley, ca. 1910.
*Courtesy of the Denver Public Library Western History Department.*

is park-and-ride country. People commonly note that they live in the town but that their environment is in the mountains.

Surface wonders anticipate subsurface riches. Rising above the plains, the cover of soil and vegetation becomes thinner, revealing the rock beneath. Geologic forces predominate, and the energetic folds, uplifts, and thrusts of the earth's forces have created a mountain landscape of great surprises and scenic splendor. Human activities parallel the changing properties of the land: in the mountains, mineral extraction; in the plains, soil formation and the tending of agriculture.

The Rockies are cut by canyons and valleys. The rock canyons are precipitous, palisaded cliffs; the views down and up are exhilarating and frightening. Sunlight never reaches the floors of some canyons. Agricultural valleys with arable land, such as the Elk or the Yampa, have rivers winding their course down the valley, crops in the floodplain, farms and ranches on the high ground, and forested slopes. Water is impounded in ponds or captured in diversion canals at the toe of the slope. Road and canal ride the edge, and their lines extend down the valley—the river, ditch, road. The sinuous courses of rivers and canals are determined by contours and drainage while the lines of communication of telephone and power lines take the path of least resistance—they are straight.

In the midst of the Colorado mountains lie four great "parks," vast open grasslands, treeless intermontane basins: North Park (elevation 8,000-9000 feet), Middle Park (7,300-9000 feet), South Park (9,000-10,000 feet), and the San Luis Valley (7,500-9,000 feet). Other, more modest parklands are found as well: the Wet Mountain Valley, Estes Park, Bergen Park, Poudre Park, Woodland Park, Egeria Park. The parks combine the characteristics of the domesticated pastoral landscapes of the English country estate and the American urban park, but at a much exploded scale. John C. Frémont called the parks "mountain coves" and noted, "The enclosure, the grass, the water, and the herds of buffalo roaming over it, naturally presented the idea of a park. Indians called North Park the "Cow Lodge"; trappers called it "New Park" and the "Buffalo Bull Pen." The Rabbit Ears, Medicine Bow, and Park ranges enclose a great basin almost forty miles across, creating a great landscape theater-in-the-round, a 360-degree landscape. The space feels circular and whole, the mountains a wainscot wrapping the base of a grand room. In the parks the

MARGARET BOURKE-WHITE  Contour Plowing, Walsh, Colorado, 1954. *Courtesy of Life Magazine,* © *1954 Time Warner Inc.*

landscape is enclosed, yet it is an "open" enclosure, as if a valley were a flower whose petals had bloomed. The sky, especially in bad weather, is palpable, a great domed lid on the landscape. The southern San Luis Valley is drier and more expansive. The valley is in the embrace of whole identifiable mountain ranges—the San Juan and Sangre de Cristo mountains, the Culebra Range and the Spanish Peaks—each distinctive in silhouette and character. The valley is large enough that one can see different types of weather from one spot, a common Colorado occurrence.

The parks are dry, harsh landscapes with hard winters and short summers. They remind one of the plains, but enclosed, the paradox of a limited spaciousness. Much of their lands and the surrounding valleys are ranch country with small timbered areas, hayfields, and grazing land. Unlike agriculture—which geometrizes the landscape in patterns of ownership, occupancy, crop diversity, and rotations—ranching creates light, thin ownership bounds, visible only when one is standing right beside them. In sparsely populated, open country, activities and structures are accentuated: a single house or windmill becomes dramatically important.

Passing through the mountains, one sees the landscape of the Western Slope slowly emerge. It is variegated terrain, as mountains, mesas, and river valleys meet. Theodore Roosevelt, on a 1908 hunting expedition near Glenwood Springs, characterized it as "great, wild country . . . the mountains crowded close together in chain, peak and tableland." The land is rawer, harsher, drier, less vegetated, a desert scrubland with woodlands of piñon pine and juniper and green-topped mesas. It is sparsely populated; newer habitation is coupled with the most ancient. The human presence is not felt for great stretches, and the human mark is often slight. The Western Slope is part of the Colorado Plateau, the domain encircling the Four Corners region. It is a complex matrix of plateaus, canyons, valleys, mesas, the occasional green of the river valley, and ranchland. It is a rugged, rocky country. Walled canyons lead to glacial valleys with ranches and farms, valleys funneling toward towns. It is a striated land, with geologic forces visible and palpable in their processes, with layering, uplift, erosion, and the stripes of color-coded formations.

It is a land of stunning and palpable surprises. The road leading to the Black Canyon of the Gunnison River climbs for miles, only to bring one to the precipice of a rock canyon gouged or sawn out of the earth. Taking switchbacks up to mesa tops, one discovers elevated tablelands with their own forests, valleys, and peaks. The mesas are aerial oases capturing winter snows in the dry landscape. The land displays the stark separation of irrigated valleys of fields and the careful grids of orchards surrounding Grand Junction, Fruita, Montrose, Delta, Olathe, and Paonia. The landscapes are two-toned: above the green valleys and over the treetops are dry white-and-tan slopes.

Booms and busts have been hard on the land. Portions have an abandoned, used-up quality, with old cabins, ranches, mines, and the ruins and remnants of ghost places. There are landscapes of extraction, where coal and oil shale have been stripped away. At Hayden Station and Craig, Cameo and Morrow Point, the stacks of immense coal-fired electrical plants reach skyward as lines buzz over the terrain. Here, at the state's western extremity, the landscape asks a fundamental question. Our relationship to the landscape is reciprocal: as resources are extracted, what are we returning to the land?

**RICHARD VAN PELT**
The Arkansas River South of Leadville, 1980. From the project and portfolio "From This Land." *Courtesy of the artist.*

# CHARACTER

*"Near one's eye ranges an infinite variety."*

CHARACTER IS A HUMAN TRAIT, the distinct stamp of individuality and personality given to the common qualities of the human species. Landscapes also have character. The character of the Colorado landscape is the product of natural processes, its distinctive past and evolving history. One can attempt to identify the elements of that character in its many forms, including its images, ideals, icons, and taste. There are many Colorado landscapes. Colorado is a land of contrasts and paradoxes, both beautiful and terrifying, peaceful and violent, quiet and powerful. The state is highly urbanized, but the landscape identity is not formed in the cities.

The infinity of colors is composed of simple building blocks, the primary colors. So too, the infinity of landscapes is composed of basic elements. In Colorado, aridity and altitude serve to accentuate primary landscape elements—rock, water, plant, and sky—and to present them in extreme and often stark relationships. The extremes were noted by early visitors, more accustomed to the continuities and gradations of eastern landscapes.

Being out-of-doors is a basic part of Colorado. Climate makes the landscape visceral, where the skin, not the eyes, is the primary mode of perception. You feel the heat of the sun or the bite of the wind on your face; winter wets and chills you to the core. Western weather changes rapidly; it is typically unpredictable. Sun-filled skies become thick with thunderclouds, gentle snows change to blizzards, a dry wash is inundated by a flash flood, sweltering heat turns to freezing temperature in hours. Captain John Bell in 1820 spoke of clouds filled with "electric fluid," and John C. Frémont remarked on entering "the storehouse of the thunderstorms." William Parsons, part of the Lawrence Party in 1858, said: "We had a

**PRIMARY ELEMENTS**

**ROBERT ADAMS**  Pawnee National Grassland, 1984. *Courtesy of the artist.*

thunder shower almost each day while we remained in the camp—and SUCH thunder as no other country ever saw. On such occasions it seemed as if the old mountain rocked to its very base. The lightning, as if let loose for holiday pastime, played among the deep gorges and rocky canons of the mountains with appalling splendor." The climate is volatile and violent: chinooks, avalanches, floods, lightning, hail, brutal sun. An 1875 summer hailstorm broke windows in railroad cars and made steel boilers look like they had smallpox. During a storm in the summer of 1990, thousands of Denver automobiles became pockmarked in a single ten-minute burst of pellets. The National Hail Research Experiment has its field headquarters located near Grover, Colorado.

Climate, like all else, changes with elevation. Moving from the plains and desert to an alpine landscape above the timberline is the equivalent of moving from the Mexican desert to the Arctic. However, elevation is tempered by orientation, exposure, and protection, creating microclimatic niches. Orientation is crucial. South-facing slopes, receiving direct sun, are drier and typically more pleasant to live on, whereas north slopes are darker and wetter and keep their snow longer. In the open landscapes of the lower elevations, summer shade is a blessing. Landscape design mitigates or takes advantage of these conditions, from trees forming an umbrella over a farmhouse to ski runs carved on forested slopes.

The Colorado sky is an active component of the landscape experience. The sky is high, dry, invigorating, clear air. Only one day in six is cloudy along the Front Range, a fact now often obscured by a brown cloud of pollution. The sky has diverse personalities: open clarity, rooflike solidity, O'Keeffe pillow puffs or crest clouds reaching across the continental divide. It is a vast sky with weather advancing from different directions, sometimes dark to one side and light to the other. The arc of the sun and the path of the weather fronts, continental movements, are perceivable. You can feel like a dweller both at the bottom of the sky and on the top of the surface of the earth. Samuel Bowles, traveling in 1868, was hyperbolic in his evaluation: "The distinctive charm is the atmosphere, so clear and pure and dry . . . invigorating every sense, softly soothing every pain, lending glory to the landscape . . . clothing every feature of nature with beauty." Recently installed paintings adorning the chamber of the House of Representatives at the state capitol are entitled *Those Colorado Skies* and *Capital City Sunlight*.

**EDWARD TANGEN** Flagstaff Mountain, 1903. *Courtesy of the Carnegie Branch Library for Local History, Boulder Historical Society Collection.*

## VISTA

Colorado is aggressively scenic and remarkably pictorial, with each bend in the road a postcard possibility. The panoramic view is fundamental to experiencing the landscape. It is a horizontal scan, similar to a panning shot in film as you pivot to take in the totality of the scene. Mountain vistas recall the foldout Swiss Alp panoramas of Baedeker guides. In fact the American western mountain sensibility was incubated in that European setting. The panoramic focuses on the distant horizon and even the line of division between earth and sky. The eastern horizon offers a flat line, whereas the multiple planes of the ranges of the Rockies offer jagged, multilayered and multileveled diorama-like horizon lines. There are over fifty named mountain ranges, many with distinct profiles, silhouettes marking the skyline. The zigzag mountain symbol is even repeated thousands of times on the state license plate. John Frémont compared the Front Range to a "dark corniced line." Stephen May wrote of Silver Mountain, "Its geometric profile arouses genuine emotions, emotions which in years past have driven perfectly normal people great distances to worship such temples."

Oftentimes the sharpness of the silhouette blurs: the distant skyline loses its sharpness, and the solidity of the mountains is confused with the ephemerality of clouds. Mountains become mirages, especially when the base is obscured and the peaks seem to float in space. In 1848, while looking west, Francis Parkman, the author of *The Oregon Trail*, noted: "I distinguished one mass darker than the rest, and of a peculiar conical form. I happened to look again, and still could see it as before. At such moments it was dimly visible, at others its outline was sharp and distinct; but while the clouds above it were shifting, changing and dissolving away, it still towered aloft in the midst of them, fixed and immovable. It must thought I, be the summit of a mountain; and yet its height staggered me. My conclusion was right, however. It was Long's Peak . . . the thickening of gloom soon hid it from view, and we never saw it again."

This is a landscape of far sight. "Nothing so much as distance informs this landscape," says N. Scott Momaday. Different and distinct landscapes can be seen, appreciated, and anticipated from great distances. There is a vertical as well as a horizontal panorama. Looking down long valleys, you can stand at the headwaters of a stream and scan an entire watershed. On a ridge of the Sangre de Cristo Mountains in 1848, George Ruxton wrote,

**DANA B. CHASE** Trinidad, Fisher's Peak in the Distance (detail), ca. 1880-1889. *Courtesy of the Denver Public Library Western History Department.* From series of stereo views captioned: "Scenes Along the Denver & Rio Grande Rail Road, The Scenic Road of America, by D. B. Chase, Landscape and Portrait Photographer. Pictures of Buros [sic], Sheep, Cattle, Round Ups, Branding, Mountains, Dale and Valley."

"From my position on the summit of the dividing ridge I had a bird's-eye view as it were, over the rugged and chaotic masses of the stupendous chain of the Rocky Mountains, and the vast deserts which stretched away from their eastern bases; while on all sides of me, broken ridges, and chasms and ravines, with masses of piled up rocks and uprooted trees, with clouds of drifting snow flying through the air, and the hurricane's roar battling through the forest at my feet, added to the wilderness of the scene, which was unrelieved by the slightest vestige of animal or human life."

The structure of much of the landscape is visible and comprehensible. F. V. Hayden noted, "No one could stand on the summit of one of these ridges and turn his eye westward over the series, rising like steps to the mountain summit, and looking eastward across the broad level plain where the smaller ridges die out in the prairies, like waves of the sea, without arriving at once to a clear conception of the plan of elevation of the Rocky Mountain range." An exuberant Samuel Bowles stated, writing of an 1868 camping expedition in the Arkansas River valley, "Every variety of scene, every change and combination of cloud and color were offered us . . . and we worshipped, as it were, at the very fountains of beauty, where every element in nature lay around, before, and above us." For explorers and travelers, the distant view equals anticipation, especially moving east to west, crossing the plains with the mountains in the distance. Bowles added his appreciation of "the magnificence of distance, of height of length, . . . which is the first and greatest most constant thought of the presence."

Places have been structured and designed to appreciate these vistas and to survey the terrain: Denver's Inspiration Point and Cranmer Park, originally named Mountain View Park, where a terrazzo mosaic identifies the two hundred miles of mountains viewed from the park; scenic drives such as Cañon City's Skyline Drive or the Trail Ridge Road of Rocky Mountain National Park with its "What you can see from here" plaques; innumerable roadside overlooks and the grand resort porches wrapping the base of Grand Lake Lodge or Estes Park's Stanley Hotel. At resorts, the porch is a place to sit and survey the scene, to observe both the social life and the view. The porch frames the landscape, and you reside in the postcard.

The distant view, especially of the mountains, represents a landscape ideal, as do the clear sky and the vast space of prairie and park. Views carry emotional weight and symbolic association. There is always an interplay between actual landscapes and ideals. The view represents promise, hope, individual aspiration, Edenic possibilities, and pure and pristine nature. "The vistas are everywhere and unending.... Here, all lines of sight are trained upon infinity," says Momaday.

The relationship to the mountains is pragmatic, symbolic, and mystical. In the western world, mountains have had dual identities. Primary forces of nature, the home of gods and clouds, they were gazed on and worshipped from below but, until the eighteenth century, rarely visited. By then mountains had become challenges for engineers, and peaks were points to be scaled. Mountains have been feared; they have been held sacred, the sites of quests. Their massiveness and scale are part of their mystique. The relationship is archetypal and as fundamental as the connection between high and low, the human plane of normal existence looking to something beyond, something spiritual, mythic, heroic, or just plain big—we are dwarfed by their presence. Isabella Bird first experienced the Rockies in a railroad car from Cheyenne. Observing the wall of mountains, she wrote, "I can look at and feel nothing else."

The Rocky Mountains, arrayed in ranges and punctuated by peaks, divide and unite the state. The Rockies are the "shining" mountains, their snow-covered and crystalline rock surfaces reflecting sunlight. They are the "mountains whose snowy peaks glitter," as John Frémont noted. For eastern Coloradans, the morning sun illuminates the mountains so that the peaks shine at first light, and the setting sun silhouettes them in the evening.

The Rockies are white-crested stone waves on a continental sea. The ridge line, the continental divide, bisects the state. For Helen Hunt Jackson, they were "like an alabaster wall rounding the very world." In their midst Horace Greeley noted, "A wilderness of mountains rose all around us, some higher, some lower, but generally very steep, with sharp narrow ridges for their summits." For Francis Parkman, they were "the whole sublime congregation of mountains."

## MOUNTAINS

ANDREA JENNISON View From the Porch, Pine Creek Ranch Near Buena Vista, 1986. *Courtesy of the artist.* This house, known as the "Westly," was ordered from the Sears and Roebuck catalogue, brought in by rail and built in 1929. The photograph was made for the Constitutional Bicentennial project "America's Uncommon Places: The Blessings of Liberty" which was sponsored by the Society for Photographic Education, Eastman Kodak Company and the National Trust for Historic Preservation.

The heights are inspiring. When Samuel Bowles and his party climbed Gray's Peak in 1868, he found it "the great sight in all our Colorado travel." He added: "No Swiss mountain view carries such majestic sweep of distance, such sublime combination of hight [sic] and breadth and depth; such uplifting into the presence of God; such dwarfing of the mortal sense, such welcome to the immortal thought. It was not beauty it was sublimity; it was not power, nor order, nor color, it was majesty; it was not part it was whole; it was not man but God.... Mountains and mountains everywhere."

Fifty-four peaks rise over 14,000 feet; they are the "Fourteeners," and to climb them all is a badge of honor among the state's mountaineers. Thirty-first in height, but best known, is Pike's Peak. Peering at the Sinai-like mountain with its bulky slopes and its oft-shrouded, oft-clouded peak, one can see why Zebulon Pike was compelled to make his foolish, hazardous, and unsuccessful trek to the summit. "Pike's Peak and its great brethren rose out of the level prairie, as if springing from the bed of an ocean," said Parkman. Pike's Peak became the gold rush symbol. "Pike's Peak or Bust" stood for the mountains, riches, and gold, and it became synonymous with Colorado. There were Pike's Peak haircuts and foods, and the mountain was pictured on fashion prints. Those who joined the rush were Pike's Peakers. The *St. Louis Evening News* in 1859 wrote: "Pike's Peak. It is magnet to the mountains, toward which every body and everything is tending... moving Pike's Peakward."

The mountains symbolize the state's changing fortunes. They stand for hope, grandeur, and limitless possibility. In 1849 Senator Thomas Hart Benton of Missouri, the great exponent of manifest destiny and proponent of a transcontinental American railroad, imagined as the rail line's "crowning honor" a mountain topped with "a colossal statue of the great Columbus" constructed "from a granite mass of a peak of the Rocky mountains overlooking the road—the mountain itself a pedestal and the statue part of the mountain—pointing with outstretched arm to the western horizon, and saying to the flying passengers, `There is the East; There is India.'" Samuel Curtis, writing from his winter quarters near Cherry Creek in November 1858, said the mountains "loom up like a steamboat in a fog." That mountain view is now often obscured, veiled, hidden, invisible behind a brown cloud of smog and pollution, largely caused by the automobile. Visions of promise have become pictures of peril.

The mountains are opposing, like an opposable thumb that holds the landscape in its grasp. They form half of a set of landscape equations: mountain-plain, mountain-valley, mountain-park, mountain-town. Along the Front Range, Frémont commented: "This mountain barrier presents itself to travellers on the plains; which sweep almost directly to its bases; an immense and comparatively smooth and grassy prairie, in very strong contrast with the black masses of timber, and the glittering snow above them. This is the picture which has been left upon my mind." Heading up the Arkansas River valley, he spoke of "the beauty of this spot, immediately at the foot of lofty mountains, beautifully timbered, which sweep closely round, shutting up the little valley in a kind of cove." Along the Huerfano in the country between the Spanish Peaks and the Sierra Mojada (Wet Mountains), Gwinn Harris Heap wrote in 1852 of the "wild and beautiful" scenery. "Mountains...towered high above us, the summits of some covered with snow, while the dense forests of dark pine clothed their sides, contrasted well with the light green of the meadows near their base." William Parsons reported in the *Lawrence Republican* on October 28, 1858: "The country around the `Peak' is beautiful beyond description. The valley in which we are encamped is hemmed in by the towering peaks of the mountain range on the west, and a line of precipitous, rocky bluffs upon the north and east."

Joseph Bijeau, a guide to the Long Expedition, told of the "country situated within the mountains" and its valleys (he was speaking of parks as well) that were without timber but that were fertile, "well watered, and adapted to civilization." Their undulating surfaces were "terminated on all sides by gentle slopes leading up to the base of the circuljacent [*sic*] mountains." Obadia Oakley's 1839 journal described coming out of South Park: "Our path was very rough and precipitous; but nestled at the bottom of the gorges, beneath frowning and overhanging crags, were beautiful little vales of the brightest verdure, enameled by a profusion of flowers rivaling the choicest specimens of floriculture, and irrigated by springs and rivulets of the most transparent and best flavored water. The contrasts between those and the rugged scenery around was pleasing in the extreme." Samuel Bowles in 1868 described South Park as "closely in the lap of the great mountains." He added, "The grand parent ranges...guard and enfold what are well called NATURAL PARKS." The parks encapsulate the state; they are enclosed plains wrapped by mountains.

**WILLIAM HENRY JACKSON**  Silverton, ca. 1880. *Courtesy of the Colorado Historical Society.*

**WILLIAM HENRY JACKSON**　Silverton, ca. 1880. *Courtesy of the Colorado Historical Society.*

**DAVID PLOWDEN**
U.S. Route 6, Loveland Pass, 1978. *Courtesy of the artist, (c.) 1982.*

Many towns have views to ranges or specific peaks: Cortez and the Sleeping Ute, Colorado Springs and Pike's Peak, Boulder and the Flatirons, Silverton and Mount Wilson, Longmont and Long's Peak, Alamosa and Mount Blanca, Carbondale and Mount Sopris, Aspen and Aspen Mountain, Walsenburg and the Spanish Peaks, Trinidad and Fisher Peak. Towns "brand" their hillsides, like the flanks of cattle, with their initials. These markers of painted stones indicate ownership and local pride. Graduating students amend the letters with annual markers. A grand *D* overlooks Del Norte, a bold *S* lords over Salida, a giant *M* signifies the Colorado School of Mines, and in Gunnison, a *W* for Western Colorado State College has superseded a more modest *G* on the opposite hillside.

## PASSES AND GATEWAYS

A litany: Buffalo, Boreas, Rabbit Ear, Raton, Muddy, Medano, Mosquito, Wagon Wheel, Loveland, La Veta, Independence, Gore, Vail, Weminuche, and dozens more. Whereas mountains are the barriers, the pass is the opening, the gateway to the other side. Marshall Sprague, in *The Great Gates: The Story of the Rocky Mountain Passes*, notes, "At the top of every gulch, in the chill air above the grass, above the timber, above the bleak moraines and lonely tarnes was a pass." Many of these high passes cross the continental divide, which separates oceanic watersheds.

Passes are the ways through the mountains, creases in the ranges, the places of going up and over. The openings were discovered, and passage was regularized by building trails and roads, and then the railroads, climbing up and snaking through. Subsequent tunneling for vehicles and water has often negated the passage over the mountains by creating an opening through the mountains. Each physical pass is also an emotional pass. There is a long ascension and exertion, the satisfaction on reaching the summit, oftentimes the first glimpse of one's destination, then anticipation, and finally the descent. The landscape mirrors the emotions; the emotions mirror the topographic experience.

Opening a pass unlatches a lock closing one side of the mountains to the other. The pass is an entry to another domain, into the mountains or the next valley. At the grandest landscape scale, the pass is a gateway, but it is echoed in the markings of boundaries throughout the landscape. Gateways mark the boundaries between domains; they are territorial markers, a welcome and a warning. Ranches and farms display the traditional

**HEYN** The Entire Population of Walden, 1887. *Courtesy of the Denver Public Library Western History Department.* "Walden in 1887. The entire population at that time. The first house on the left was built by I.H. Greene in 1885, the first house in Walden. 1. Ruth Rathbun 2. Mrs. Shippey 3. Maud Rathbun 4. Ike Greene (with transit) 5. Doc Squires (a veterinarian) 6. A Rathbun girl 7. Mrs. Rathbun 8&9. small Rathbun children 10. Austin Greene 11. John Shippey with his daughter Alice in his arms 13.[sic.] Mrs. Ike Greene 14. Valdai Shippey 15. Iva Greene 16. Tim Greene 17. Jack Greene" Caption from back of print.

wooden arches, crowned with names, brands, and skulls. There are the grander arches built to announce the passage to Denver's mountain parks or, outside of Union Station, the city's great "Mizpah" welcome arch that greeted travelers from 1906 until its demolition in 1931. New suburban developments mistakenly eschew these classic statements and seek distinctive and singular "entry statements."

## FRONTIER

The history of the land resides not only in its relict markings, patterns, and organization but also in its fundamental formative ideas. The spirit of the early periods of Colorado's development perseveres. In many ways Colorado remains a frontier landscape. Frederick Jackson Turner's thesis on the American frontier was postulated a century ago. The concept of the frontier as cultural determinant is still debated, but it remains a powerful idea. The frontier is a spatial concept of an advancing edge, a cultural "front." In the nineteenth century, the eastern side was perceived as an inexorably advancing line of "civilization" entering a "savage" wilderness. Seen from one direction, the advancing edge was progress, an inevitable and desirable movement. From the opposite direction, on the other side of the line, it was an invasion. From the distant space of history and anthropology, it was a meeting ground of cultures.

The Colorado frontier landscape persists, but what is a frontier landscape? The boundary zone is a place of confrontation. The frontier is a new, raw, vibrant, often violent place. Free, unfettered, and unregulated, the landscape has these qualities, the vaunted values of the American individualist ethic. But there are also countervailing tendencies, the cooperative endeavors of trail blazing, building, cooperatives, ditch digging, and community organizations. The edge is also a place where both sides can be seen. You are not in the midst but at the periphery, where one can straddle the line and look across divisions. It is a place of choices, a dichotomous circumstance. The edge is fragile and unstable and temporary, with the sides rarely in balance, for one side typically seeks dominance over the other.

Sometimes, one can see the frontier as an advancing front in the landscape, as a true and sharp boundary, with clear lines, barriers, edges, and divisions experienced on the land. As suburban development enters a rural landscape, lawns are found on one side of the road while crops or rangelands are on the other. The boundary is not always neat. There is also

(Left) **L. C. MCCLURE**  Sangre de Cristo Range from Silver Cliff, ca. 1900. *Courtesy of the Denver Public Library Western History Department.*   (Above) **KENNETH HELPHAND**  Naturita, 1989. *Courtesy of the author.*

the leapfrogged edge, of pioneers and elements extended beyond their range, from the lone homesteader to new exurban communities. These cultural artifacts recall pioneer plants in their first step toward naturalization.

Even today there is a boundary between an urban, civilized, and technological landscape and a wild landscape. Only now the values have been inverted, and virtue is often seen to reside on the side of the wild, with landscape planning and design seeking to codify the edge with wilderness areas and parks. Coloradans and westerners want both, so mountain resorts offer natural adventure along with civilized, even decadent, pleasures. This desire for the best of both worlds is not a new ideal but is part of a historical continuum, the ancient Roman ideals of a primeval pair of city and country resurfacing in modern guise. Thus mining towns had opera houses; contemporary communities have gourmet restaurants and trendy shops—all symbols of taste and urban culture. On a more modest and civic scale, the schoolhouse, courthouse, and post office once signified civility. Even the modern city dweller seeks the best of both worlds, and suburban lots are advertised as having both mountain views and proximity to downtown amenities.

**E.K. EDWARDS** Residential Street Construction, ca. 1950–1960. *Courtesy of the Colorado Historical Society.*

## BOOM AND BUST

The continuing cycles of boom and bust that have characterized Colorado's economic fortunes are visible in the landscape. Booms are times of growth, optimism, building, and abundant resources and energies, times when fortunes are made and bright futures imagined. Just as plants are tropic, growing in the direction of light and water, so too booms find people, capital, and attention leaning in their direction. Busts are times of stagnation, pessimism, minimal resources and energies, misfortune, and disillusionment. Busts are similarly tropic: everything seems to turn away from the direction of growth, and places seem to wilt.

What do booms and busts look like? A boom landscape is vibrant, chaotic, unplanned, and evolving, a time of rapid, hasty building. It is a landscape under construction, moving up hillsides, filling in vacant lots, creating skylines, squandering resources, and thinking little of tomorrow. Booms have few vacancies; they keep sign makers busy and speculators active. Busts are periods of slow decline, with landscapes neglected, abandoned, even in states of

ANDREW JAMES HARLAN  Victor After the Fire, 1899. *Courtesy of the Cripple Creek District Museum.*

**ANDREW JAMES HARLAN**  Victor Ten Days After the Fire, From Ajax Mine, August 1899. *Courtesy of the Cripple Creek District Museum.* In this view, looking the opposite direction from the previous photograph, canvas tents and wooden buildings have sprung up as the town was rapidly rebuilt.

**ARTHUR ROTHSTEIN**   Farm Abandoned Because of Continuous Crop Failures, Weld County, October 1939. *Courtesy of the Library of Congress, Farm Security Administration Collection.*

devolution. In busts, mistakes are reflected on, vows are made to do things differently, lessons go unlearned, and the next boom is awaited.

Colorado has experienced boom-and-bust cycles in mining, agriculture, ranching, energy, technology, and even snow. The fate of the landscape has risen and fallen with these fortunes and misfortunes. However, the pace and the pattern of landscape change differ from the precipitous peaks and deep valleys of the economic cycle. Short-lived building booms leave relics that last for years, whether they be in a high-country mining town or Denver's skyline. For other elements, the landscape changes more quickly in its adaptation to the economy. Bust signs are easily read: fallow fields, vacant offices, empty shops, closed mines, "for sale" signs, stalled projects, idle construction sites.

## GHOST LANDSCAPES

Landscapes are time lines where the patterns of change and development are recorded. Following a trail or road can take us through a series of "time zones," places that embody particular moments of the past. In Colorado the time line ranges from the almost incomprehensible processes of geologic transformation to the relics of ancient cultures, to almost instant landscape changes. Historical processes of growth-development-decay seem to be accelerated. The contemporary landscape is one of future shock, where instead of the gradual accretion of generations, we find a quickened pace. Thus old-timers can recall the settlement, boom, bust, and abandonment of areas. People remember Aspen, Telluride, and Breckenridge as mining towns; the energy boom of the West Slope caused towns to double and triple in size, only to contract a few years later. Places become like actors, quickly costumed to play new roles.

We know of ghost towns, but there are also ghost landscapes. Ghost towns are easily recognized by collapsed buildings, relict structures, and rusting machinery. In their spaces we sense a tactile, visceral connection to the historical, human presence; we experience sensations of nostalgia, from melancholy to sentimentality. We are less attuned to the framework and setting of what can be described as a ghost landscape. We are less aware of the physical clues to a corresponding emotional expression.

All landscapes are in a sense ghost landscapes, relict landscapes of previous inhabitants. There is a normal process of decay, as nature reasserts her position, but in arid and high-

country landscapes, human impacts scar the land, creating permanent indicators of long-lost activity. Some cultures, such as the Utes, the Arapahos, or the Cheyennes, left few physical remnants, whereas activities such as mining made marks that will persist in perpetuity. Thus, our history is skewed by what remains, what we see, what is accessible, and what we focus on.

Colorado is a modern landscape. It was largely settled by Euro-Americans in the industrial era, most of it in modern times. It is not a traditional space—modern imperatives and phenomena are present here in "purer" form. For example, the railroad did not have to pass through built landscapes or cultivated fields but could assert its own technological imperative. With only a few exceptions, the relict landscape existing from previous inhabitants has largely been obliterated or ignored. Most of what happens on the land is happening first; it is not layered on top of a previous cultural landscape (or not perceived to be so). Tourism, a quintessential modern activity, has been part of the landscape since almost the state's beginning, and virtually all of the state's history since the 1850s is documented photographically.

Kevin Lynch has likened the landscape to a temporal collage, a haphazard but artful series of overlays and revelations. Redstone, an experimental company town, died but was resurrected with second homes, antique shops, and historic preservation. Mining camps are reborn as resorts. One generation's landscape of difficulty becomes another's adventure in four-wheel drive. Times, as well as spaces, exist side by side. The historian Robert Athearn has written of the towns he knew in the 1920s, "It was almost as if the towns existed in one century and the surrounding countryside in another, the degree of difference depending on how far one lived from those urban centers." There are places where, with a modest act of historical imagination, you feel as if the landscape is naturally populated by its early inhabitants—mountain men, explorers, miners, or cattle drovers. The space takes you to another time.

**PETER GOIN** Abandoned Road, 1981. *Courtesy of the artist.* "This road was once a stagecoach route deserted in the 1930's when a wider, more direct road was completed. The bridge is over a dry creek." Artist's caption.

Colorado—the name itself means "colored red," after the red ocher rivers of the region. Colorado has a palette that changes as one moves in space east to west, north to south, and vertically. The spectrum changes with the seasons and the weather. Streams turn from an

## SUBLIME AND BEAUTIFUL

**JERRY UELSMANN** 1971. *Courtesy of the artist, (c.) 1971.* Visual artists, especially photographers, have long been drawn to the spectacular beauty and rich history of the Colorado Rockies. Here, Uelsmann uses the abandoned mining town of Independence as a backdrop for this self-portrait collage as he literally and figuratively walks back into history.

icy gray to red, yellow, or brown as rains bring soil down from the uplands. In the 1930s the Colorado Association, a booster organization, proclaimed the state to be "Colorful Colorado."

Wallace Stegner has noted that when one moves up the Platte Valley, green ceases to be the prevailing color of the earth and "tans, grays, rusty reds and toned white" predominate. For the photographer Robert Adams, the plains are a "dead, light brown." Crossing the Palmer divide in the summer of 1845, John C. Frémont found "the whole valley . . . radiant with flowers; blue, yellow, pink, white, scarlet, and purple, vied with each other in splendor. . . . Crossing to the waters of the Platte, fields of blue flax added to the magnificence of this mountain garden."

The spectrum of the earth is tilted toward red, the rusted evidence of iron in the rock and soil—Red Rocks, Redstone, Redmesa, Red Cliff, the Garden of the Gods. Along the Front Range, Susan Tweit saw "a red valley, the colors of rock ranging from soft salmons to brilliant rusty orange red." F. V. Hayden described the sandstone beds of the West Plum Creek valley as "quite variegated, of almost every color and texture, mostly fine sand, brick red, deep yellow, rusty red, white ash colored, dull black, etc." Zane Grey saw the Flattop Mountains as "dark green slopes of alpine spruce, rising to bare grey cliffs and domes, spotted with white banks of snow." Samuel Bowles stated, "The prevailing tone and impression of the [Middle] park is a coldish-grey. You find it in the earth; you see it in the subdued, tempered or faded greens of leaf and shrub and grass; it hangs over the distant mountains; it prevails in the rocks; you feel it in the air." In fall the mountains are banded: at top are the white peaks, then the aspens that have turned a golden yellow-orange color that is so striking it should be called "aspen," and finally dark spruce. Plants in arid landscapes have a limited palette and tone: silvery blue, gray-green, mottled greens, sage-green. The colors connect us to other senses, the recollection of the smell of sage after a rain.

In winter the land is browns, tans, and grays, with the equalizing blanket of winter white. The white reflects, glistens, blinds, and during full moons, helps blur the boundaries between night and day. Isabella Bird described an eastern sky that successively changed from chrysoprase to aquamarine to emerald, the colors of three gems. "Unless I am color-blind, this is true," she swore. Walt Whitman looked to the sky and its "atmospheric hues . . . aerial

gradations and sky effects inimitable, nowhere else such perspectives, such transparent lilacs and greys." He continued: "I can conceive of some superior landscape painter, some fine colorist, after sketching awhile out here, discarding all his previous work . . . as muddy, raw and artificial. Near one's eye ranges an infinite variety; high up the bare whitey-brown, above timber line; in certain spots afar patches of snow any time of year. . . . As I write I see the Snowy Range through the blue mist, beautiful and far off. I plainly see the patches of snow."

There is a more fundamental aesthetic. On February 5, 1807, Zebulon Pike wrote: "We ascended a high hill, which lay south of our camp, from whence we had a view of all the prairie and rivers to the north of us; it was at the same time one of the most sublime and beautiful inland prospects ever presented to the eyes of man. . . . In short, this view combined the sublime and the beautiful; the great and lofty mountains covered with eternal snows, seemed to surround the luxuriant vale, crowned with perennial flowers, like a terrestrial paradise."

Pike seems to have carried books, along with an aesthetic, in his head. This western explorer looked to the landscape through the lens of eighteenth-century theory, which is still part of our landscape vocabulary. In 1757 the English philosopher and statesman Edmund Burke wrote *A Philosophical Enquiry into the Origin of Our Ideas of the Sublime and the Beautiful*. As Burke explained, the sublime and the beautiful are polar opposites. The sublime is an emotion based ultimately on fear and its characteristics. In the landscape it evokes wonder, awe, power, vastness, difficulty, infinity, and magnificence. These are clearly embodied in the scale and immensity of the "white mountains" of which Pike speaks and also in Colorado's mesas, canyons, plains, panoramas, sky, thunderstorms, snows, and torrential streams, which are all sublime. The beautiful, on the other hand, is an emotion founded on pleasure and is manifested in smoothness and curves. Often it is discovered in details. It is surely found in the fecundity of river valleys, the sinuous lines of streams, the shimmer of aspens, the curve of a columbine. It was the *combination* and *contrast* of these characteristics that so impressed Pike. It is this duality that dramatizes the Colorado landscape, especially in the parks, the great beautiful "hidden valleys" ringed by sublime peaks. One hundred and twenty years after Pike, Courtney Ripley Cooper spoke of the Colorado high country in similar terms, finding it "immutable, mysterious, grudging of invasion; treacherous; petulant, yet calmly loveable and alluring."

**LAURA GILPIN**  Landscape Class, Broadmoor Art Academy, 1920. *Courtesy of the Amon Carter Museum, Fort Worth, Laura Gilpin Collection.*

**WILLIAM HENRY JACKSON** Quandary Peak and Blue River Range, 1873. *Courtesy of the United States Geological Survey, Denver.* Plate #82, from a 5-part panorama made by Jackson from Mt. Lincoln near Breckenridge.

**WILLIAM HENRY JACKSON**  Gray's Peak and Hoosier Pass, 1873. *Courtesy of the United States Geological Survey, Denver.* Plate #83 from the 5-part panorama taken from Mt. Lincoln.

(Left) **WILLIAM HENRY JACKSON**  Pikes Peak from the Gateway, Garden of the Gods, ca. 1880. *Courtesy of the Colorado Historical Society.*  (Above) **ROBERT G. ZELLERS**  Cresson Mine, Cripple Creek, 1942–1946. Courtesy of the Colorado Historical Society.

The vocabulary may be archaic, but the emotional responses are not. Sublime and beautiful, fear and pleasure, still typify basic responses to the Colorado landscape, only now it is human effects that are paramount. New elements have been injected into the equation, and sadly these are largely of fear. Dams are awesome, but we know the catastrophe of collapse. Modern agriculture is a marvel, but we still remember the dust bowl. Although the grand panorama of the mountains remains, it is seen at times through the obscure film of brown haze. At the extreme are the horrors of mutation and apocalypse, a landscape made toxic, poison places of waste heaps, Rocky Flats, the Climax mine. Distance is no longer a comfort, the scenic dimension (and delusion—out of sight, out of mind) is no longer an amelioration, and removal to a viewpoint is no longer a protection.

## MINING METAPHOR

Each landscape and land use generates its own vocabulary, which oftentimes can be applied to other circumstances. As a form of landscape metaphor, these vocabularies are more than wordplay, for there is the close kinship of a common setting and the similarity of emotional response. Mining is one of the state's master metaphors. Mining was destined to be the basis of the state's economy until after the turn of the twentieth century, but mining terms have been used to illuminate agriculture, describe urban development, and help explain tourism. The metaphors can be particularly illuminating in a setting where the landscape and the activity have changed. The older vocabulary still carries meaning. Colorado is still "mined," the "gold rush" was only the first of many, and "tailings" are left everywhere as a consequence of resource exploitation.

Character can also be defined negatively, by what it isn't. What isn't "Colorado"? There are many responses. It isn't gray skies, "eastern," resort towns, the new downtown Denver. Some of these qualities are clearly inappropriate or out of place, but others are simply ways of expressing concern for the undesirable effects of development. Current residents will assert that "easternization" began in the sixties. Historians will agree that it did—in the 1860s—for with the exception of the Indians, most settlers began as easterners. Eastern power, embodied in the federal government, has made Denver its western capital.

# EXPLORATION

*"Here is everything that a good settlement needs."*

Exploration is the process of learning a landscape. People come to a landscape with a set of expectations founded on their previous knowledge, experience, dreams, and desires. They may be on a quest, sacred or profane, or they may be just seeking a passage. Exploration is a formative act, but it is an activity that continues, thus there is something that unites nomadic plains-dwellers, explorers, mountain men, miners, settlers, travelers, and tourists. All seek something from the land, respond to the land as they experience it, and modify the landscape for themselves and future explorers. They change it through their activities and also through their accounts, which set the expectations and form the groundwork for the next visitor or settler.

The exploratory process has stages. One enters a new place with a set of predilections and expectations. For Colorado, these have been mythic ideals: the accounts of those who went before; the willful exaggerations of the promoter, the booster, and the huckster; the lure of the tourist brochure, the photograph, and the movie. First impressions linger. They become the touchstone of remembrance: the farmer who recalls a field before its domestication; the resort owner who imagines gondolas climbing mountainsides. Explorers guide others to the land, but explorers are not the settlers, those who by choice or accident will create the landscape.

The explorer not only is the first but also is a primary and formative influence. There is a primacy of first steps in the landscape—in seeking pathways, in surveying, mapping, and

## QUESTS AND QUESTIONS

**Photographer Unidentified** The Continental Divide Seen From Doany Hill, Jefferson County, June 2, 1918. *Courtesy of the Colorado Historical Society, State Highway Department Collection.*

platting the land, in selecting sites for mines, farms, rails, roads, and settlements. All of these weigh heavy and set the frame and preconditions for the future.

In the exploratory process there are the immediate needs of wayfinding, orienting oneself in the landscape, choosing the right path, and deciding which advice to heed or which map to follow. There is also the necessity of monitoring the condition of individuals, animals, vehicles, resources, and materiel. But each journey has a mission: to look for information, routes, or sites and to evaluate the landscape by those criteria as it is traversed, seen, documented, and collected. The accounts of these journeys can be read as the explorers mission statements.

Colorado's first European explorers were Spaniards moving north from what is now New Mexico. The first written account is the diary of Fray Silvestre Velez de Escalante, who was on an expedition led by his superior, Fray Francisco Atanasio Domínguez. They crossed the Colorado Plateau going south to north in the summer of 1776, following trails used by Spanish traders, who followed Ute trails on the south and west sides of the Sierra de las Grullas (Cranes), now called the San Juan Mountains.

The diary of the Domínguez-Escalante Expedition clearly reveals what the Spaniards were seeking. Their mission was to reconnoiter the land for potential settlement sites. "Here is everything that a good settlement needs for its establishment and maintainence as regards irrigable lands, pasturage, timber and firewood." Escalante noted, "Here it has a very large meadow, which we named San Antonio, of very good land for farming with the help of irrigation, together with all the rest that a settlement requires by way of firewood, stone, timber, and pastures—and all close by." This site was west of contemporary Caracas and is now irrigated farmland.

They were not blind to other aspects of the land. From the top of a ridge near Horsefly Creek, Escalante observed, "It is an eminence of very good pastures and of pleasant scenery due to its thickets of beautiful poplar groves, briefly spaced from one another." The Spaniards marveled at beavers, which "constructed ponds so big that they [resembled] a more than medium sized river at first sight." And the members of the expedition were remarkably perceptive about the relationship between seeing the land and taking possession of it. Escalante noted the Indians' worry that the Spaniards were "scouts intending to conquer the land."

In 1806 Zebulon Montgomery Pike was sent by President Thomas Jefferson to find the headwaters of the Red and the Arkansas rivers and to assess Spanish strength in the region. Jefferson, who never went west, had a distant, but significant, effect on the Colorado landscape. He sent Meriwether Lewis and William Clark on their explorations, purchased the Louisiana Territory in 1803 (the eastern half of the state is from these lands), and designed the Ordinance of 1784, thereby putting the surveyed stamp of squares of township, section, and range across the American landscape. For three years, from 1858 until the Colorado Territory was established in 1861, the area of what had been western Kansas and Nebraska territories was the Territory of Jefferson.

Pike's journals, reports, and maps reveal fundamental responses to the Colorado landscape: the contrast between the eastern plains and the river valleys, the encounter with the mountains from first view to the drama of ascending the peaks, and the implicit aesthetic that all travelers and residents carry in their minds.

## FIRST VIEW

At 2:00 P.M. on November 15, 1806, Pike saw the mountains 120 miles in the distance. Nowhere in the East is such a distant vista possible. "I could distinguish a mountain to our right, which appeared like a small blue cloud; viewed it with a spy glass, and was still more confirmed in my conjecture. . . . When our small party arrived on the hill they with one accord gave three *cheers* to the *Mexican mountains*. Their appearance can easily be imagined by those who have crossed the Allegheny; but their sides were whiter as if covered with snow, or a white stone. Those were a spur of the grand western chain of mountains, which divide the waters of the Pacific from those of the Atlantic oceans."

The wonder of the initial view and its mirage-like appearance are echoed by virtually all accounts of those who have approached the state from the east. It is one of the quintessential westering experiences. Captain John Bell, the official journalist of the Long Expedition of 1820, wrote: "We discovered a blue strip, close in with the horizon to the west—which was by some pronounced to be no more than a cloud—by others to be the Rocky Mountains. . . . The whole range had a beautiful and sublime appearance to us, after having been so long confined to the dull uninteresting monotony of prairie country." In Dr. Edwin James's account of the same journey, he noted, "For some time we were unable to decide whether

**WILLIAM HENRY JACKSON** North From Berthoud Pass (Harry Yount, explorer and first ranger of Yellowstone National Park), 1874. *Courtesy of the Colorado Historical Society.*

what we saw were mountains or banks of cumulus clouds skirting the horizon, and glittering in the reflected rays of the sun." The mountains "became visible by detaching themselves from the sky beyond, and not by emerging from beneath the sensible horizon.... The effect was several times so perfect and beautiful as to deceive almost every one of our party." In his 1839 journal, Obadia Oakley, en route to Oregon, wrote, "The Rocky Mountains were now in sight, apparently but a little distance off, swelling upon the vision in indescribable grandeur and magnificence—the most imposing scene I ever beheld; their hoary and snow-crowned summits lifting themselves towards heaven." Along the Kansas Pacific Railroad, parallel to modern U.S. Highway 50, were a succession of stops. The names are instructive: Monotony, the last Kansas station, Arapahoe, Cheyenne Wells, and twenty-seven miles from the border, the town of Firstview, so named for its distant mountain vista.

Nine days after this first sighting Pike, along with three others, began the ascent of the "Grand Peak," probably the "small blue cloud," which was later to bear his name. "We commenced ascending, found it very difficult, being obliged to climb up rocks, sometimes almost perpendicular.... We encamped in a cave, without blankets, victuals or water." The next morning they were hungry and sore "but were amply compensated for toil by the sublimity of the prospects below. The unbounded prairie was overhung with clouds, which appeared like the ocean in the storm; wave piled on wave and foaming, whilst the sky was perfectly clear where [they] were." They climbed but could get a view only to Grand Peak. Pike believed "no human being could have ascended its pinical [sic]." Now thousands do so on summer days by car, cog railway, and on foot.

In the summer of 1820, Major Stephen Long, of the United States Topographical Engineers, led an expedition up the Platte and then south along the base of the Rockies, returning east along the Arkansas River. The group of twenty-five included an official journalist (Captain John R. Bell), a naturalist (Dr. Edwin James), a zoologist, a landscape painter (Samuel Seymour), guides, interpreters, hunters, soldiers, and topographers. The expedition is best known for Dr. James's ascension of Pike's Peak and Long's evaluation of the plains and prairies that they passed through en route to the mountains. Long's maps would label the area "The Great American Desert," a land without wood or water, "unfit for cultivation, and of course uninhabitable by people depending upon agriculture for their

**WILLIAM HENRY JACKSON** View Across Blue River From Near Ute Peak (detail), 1874. *Courtesy of the Denver Public Library Western History Department.* From stereo view series: "Views Among the Rocky Mountains of Colorado. Expedition of 1874. William Henry Jackson, Photographer. Department of the Interior, U. S. Geological Survey of the Territories. F. V. Hayden, U. S. Geologist, in charge." While 19th-century survey photographers used glass negatives ranging in size from 5" x 7" to mammoth 20" x 24" Plates, they also produced stereo views of many of the sites. These three-dimensional photographs were widely distributed on the East Coast and had broad appeal to the lay public. This image shows an expedition member leaning on Jackson's portable darktent used to prepare and process his glass Plate negatives. The other member writes or sketches in his notebook.

subsistence." He likened the land to the deserts of Arabia. This viewpoint, based largely on treelessness as an indicator of arability, ends only with the successful introduction of a new agricultural technology based on the steel plow, the windmill, and barbed wire.

Despite Long's conclusion, the accounts of both Bell and James present a more complex view of the plains. This place of "hopeless and irreclaimable sterility" is also described as "verdant" and as the home to a fecundity of wildlife: deer, prairie dogs, antelope, hare, beaver, fox, elk, wolf, lizards, snakes, bear, eagles, and of course bison. The expedition members looked out across meadows filled with herds of buffalos—like "columns of a large army"—"emerging from between the distant swells of the prairie" and "blackening the whole surface of the country." James anticipated the limits of exploitation and the devastation of this species. "It would be highly desirable that some law for the preservation of game might be extended to, and rigidly enforced in the country where the bison is still met with; that the wanton destruction of these valuable animals, by the white hunters, might be checked or prevented."

The expedition traveled along the parallel ridges of the "immense rampart" of the Front Range. Along this route, where most Coloradans now dwell, east-west references are paramount. The explorers climbed a rampart and overlooked mountains in one direction. "To the east, over the tops of a few inferior elevations, lay expanded the vast interminable prairie one which had long held our mountainous march. The undulations which swell its surface now disappeared, and the whole lay like a map before the observer." Their eyes followed the course of the Platte to their camp, where their vision was comforted by the fact that the camp was "the only spot in this boundless landscape where the eye could rest on the work of human hands." Climbing the same ridge in the next century viewers, would find solace in the opposite, the surviving remnants of native landscape.

They were impressed by the picturesque and romantic character of the hogbacks before proceeding up the Arkansas to Royal Gorge. Near the site of Cañon City, Bell wrote of camping "surrounded by the grandest and most romantic scenery. . . . What a field is here for the naturalist, the mineralogist, chemist, geologist and landscape painter." They then explored the Pike's Peak region. James, the naturalist, was sensitive to both the grand scene and the detail and was the first to ascend the mountain. Above the "defined lines, encircling

the peak" was "a region of astonishing beauty, and of great interest on account of its production." Writing in mid-July, he was witness to wildflower displays where the coloring was "astonishingly brilliant." From the peak he looked to "innumerable mountains" and Indian encampments, and he traced the course of the Arkansas River "as on a map, by the line of timber along their courses." It was 42 degrees at the summit, 96 at the expedition's camp on the plains.

While exploring a tributary of the Purgatory River, a place of "bareness" and "desolation," James was "often surprised by the occurrence of splendid and interesting productions springing up . . . in situations that seemed to promise nothing but the most cheerless and unvaried sterility. Operating with unbounded energy in every situation, adapting itself with wonderful versatility to all combination of circumstances." Emerging at the mouth of the Purgatory, he wrote of Colorado's contrasts:

> Here the interminable expanse of the grassy desert suddenly bursts upon our view. The change was truly grateful. Instead of a narrow crooked avenue, hedged in by impending cliffs and frightful precipices, a boundless and varied landscape lay spread before us. The broad valley of the Arkansas, studded with little groves of timber, and terminated in the back ground by the shining summit of James' Peak, and numerous spurs of the Rocky Mountains, with the snowy pinnacles of the more distant ranges . . . before us, lay the extended plain, diversified with vast conic mounds, and insulated table-like hills, while herds of bison, antelopes and wild horses, gave life and cheerfulness to the scene.

In 1842 and again the following year, John C. Frémont led an expedition through the Rockies to survey routes west. The expedition had a political as well as a scientific motivation, for Frémont's father-in-law was Senator Thomas Hart Benton, the manifest destiny advocate. Kit Carson, later to become a western legend, acted as a guide; also along on the trip was Charles Preuss, a German cartographer who recorded the landscape in maps and sketches. A new daguerreotype camera was brought along, but it and the photographs were lost when an experimental inflatable raft was destroyed in a rapid.

**LAURA GILPIN**  South Park, 1941. *Courtesy of the Amon Carter Museum, Fort Worth, Laura Gilpin Collection.*

## MAPS

**L. C. MCCLURE**  Summit of Treasure Mountain, L. C. McClure photographing in the Elk Range, Gunnison County, ca. 1900. *Courtesy of the Denver Public Library Western History Department.*

**OLA ANFENSON GARRISON** A Good Factory Site [near DeBeque], ca. 1913 - 1915. *Courtesy of the Colorado Historical Society, Rio Grande Collection.* Published in *Stored Energy - Billions of Barrels of Oil*, Standard Shale Distillation Company pamphlet.

Ascending the Platte Valley in July, Preuss said it "looked like a garden, so rich was the verdure of the grasses, and so luxuriant the bloom of abundant flowers." But like most arriving from the East, he was struck by the noticeable lack of trees. "Timber is so scare that we generally made our fires of bois de vache. With the exception of now and then an isolated tree or two, standing like a lighthouse on the riverbank, there is none whatsoever to be seen." Frémont likened camping along the snowmelt-swollen Platte to pitching tents on the seashore.

Frémont was particularly curious about the mountain parks and changed his planned course to visit them because "the coves, the heads of the rivers, the approximations of their waters, the practicability of the mountain passes, and the locality of the THREE PARKS, were all objects of interest, were unknown to science and history." To reveal this land to science and history was his expedition's mission, and his descriptions reveal his criteria for landscape evaluation. A beautiful landscape combined the practicalities of ample water, grass, and game and pleasant weather—the necessities of settlement—with scenic delight. Of the "elevated cove" that is the source of the Platte, he said, "Certainly no river could ask a more beautiful origin."

Writing in 1848, the Englishman George Ruxton combined many of the explorers' contrasting and ambivalent responses. He was impressed by the mountains but found "the perfect solitude of this vast wilderness" to be "most appalling" while the vast prairies were "a sea of seeming barrenness, vast and dismal." Along Greenhorn Creek, he both evaluated and prognosticated. "This valley will, I have not doubt, become one day a thriving settlement, the soil being exceedingly rich and admirably adapted to the growth of all kinds of grain."

Ruxton also gave due credit to the mountain men who had preceded the organized expeditions. "All this vast country, but for the daring enterprise of these men, would be now a *terra incognita* to geographers, as indeed a great portion still is; but there is not an acre that has not been passed and repassed by the trappers on their perilous excursions, the mountains and streams still retain the names assigned to them by the rude hunters, and these alone are the hardy pioneers who have paved the way for the settlement of the western country."

The contrast many spoke of could also be viewed as the best of both worlds. This too is a continuing Colorado theme. Trappers believed that the upper and the lower worlds united at sites such as Medicine Springs (Manitou Springs). Here "his satan's majesty comes up from his kitchen to breathe the sweet fresh air." Ruxton noted: "Never was there such a paradise for hunters as this lone and lovely spot. The shelving prairie . . . is entirely surrounded by rugged mountains, and containing perhaps two or three acres of excellent grass . . . immediately overhead is Pike's Peak . . . ridge after ridge, clothed with pine and cedar, rises and meets the stupendous mass of mountains, well called 'Rocky.'"

Gwinn Harris Heap was the journalist for Lieutenant E. F. Beale's expedition of 1853. Both the scenery and the fecundity of the landscape impressed him. In alpine valleys and intervales he found "sylvan paradises" with rich pasturelands, arable soils, abundant waters, and timber. He envisioned the future and imagined valleys "unequalled in loveliness and richness of vegetation" as attracting immigrants and being transformed into "a continuous line of farms." In the foothills of the "Sahwatch" range, he wrote, "All around us the hills and the mountains were covered with rich verdure; beautiful copses and crops of trees diversified the scenery, giving it the appearance of a settled country, only wanting dwellings to rende it a perfect picture of rural beauty."

By the time Ferdinand Hayden led a series of U.S. Geological and Geographical Surveys of the Territories (1869-78), the predictions of earlier expeditions had proven correct. Around Monument Creek he found valleys "occupied by farmers and stock raisers." He noted, "Almost every available spot is taken up by actual settlers." In places, the diversity of the developing landscape was visible in microcosm. Atop a 300-foot-high dike at Valmont, just east of Boulder, he looked up and down the valley. He commented that "within a single scope of vision," the abundance and variety of Colorado resources could be seen: railroads in operation, fields, and hills with numerous openings for coal" (i.e., transportation, agriculture, and mining). The cycle of settlement had been compressed as areas had, in less than a generation, experienced the boom and the bust. "Granite bid fair to become a city and spread itself over a considerable area on both sides of the mines. Now the buildings are all fast going to decay. A few miners linger among the old 'placers,' but the mining period has passed away," wrote Hayden.

Traveling under the order of the secretary of interior to survey the geological, mineralogical, and agricultural resources of the territory, Hayden's team were mobile scientific field stations and included topographer, artist, entomologist, botanist, mining engineer, metallurgist, zoologist, cook, teamsters, laborers, and the photographer William Henry Jackson. Science was allied with a new art. Hayden, in speaking of Jackson's panoramic photographs, wrote, "Especial attention has been paid, all the time to make these views instructive as well as pleasing to the eye."

Pike followed corridors that are still used today. His crude but beautiful maps are landscape pictograms, marking essential elements. Rivers are curved lines, scallops mark the terraces of flood plains, mountains are rounded or peaked, while other symbols mark "prairie nubs." The maps demonstrate a perception of the landscape confined and limited by the corridor route, a narrow band or river valley opening up only in prairies and parks, with views toward distant pinnacles. Vast spaces were left blank. The experience was much like modern driving, in which the road corridor channels and directs the traveler's landscape experience.

In the summers of 1873, 1874, and 1875, the Hayden surveys mapped sixty-nine thousand square miles of the Colorado Rockies. The survey was completed in 1876, the year of statehood, and in 1877 the *Geological and Geographical Atlas of Colorado and Portions of Adjoining Territory* was published. These maps were to serve as the base map for the state, marking the ground on which subsequent development would occur. A remarkable document, this work was the culminating act of transforming a terra incognita into a known landscape. It contains no blank spaces: in a dizzying pattern, the contours are all filled in, river courses are all marked, and the routes of railroads and wagon, which had already penetrated well into the mountains, are all shown. Numerous proposed routes are also demarcated.

Hayden's teams surveyed the territory, set up topographical stations at commanding points, including ninety-six numbered stations (mostly peaks), twenty-five to thirty minor ones, and all important passes. From these stations, they made sketches of whole drainage basins, measured peaks, and created the base for their geological maps. Cairns they erected still survive. According to the historian Marshall Sprague, "They named everything that had no name." They measured and collected samples and images of everything: fossils, minerals, meteorological readings, water flow and streams, skeletons and shells, plants and animal samples, even leeches.

The economic map in the atlas was both a measure of progress and an anticipation of possibilities. Pinks and beiges showed the river valleys as agricultural land, the plains, prairies, and parks as pasture, and the sage and badlands on the Western Slope. Greens and blues categorized the forests by pines, piñon and cedar, and quaking aspens. Areas above the timberline were a white spine of ridgelines. Underlying this data, the primary mission of the survey was clearly presented and elaborated on in six detailed maps and accompanying geological sections highlighting coal lands, gold districts, and silver deposits. Old-timers looked for readily discernible surficial signs in the landscape, but geological mapping was the essential aid for a scientific quest for minerals. For the modern prospector, the maps were guides to wealth.

# PART TWO

*Rocks*

*Water*

*Plants*

# ROCKS

## *"The country turned inside out."*

IN ITS PRIMEVAL FORM, the eastern United States was blanketed by a temperate forest gradually giving way to the grasslands of plains and prairie, which ran to the base of the Rocky Mountains. There, the geological structure of the land was and is paramount, and it is the formations of rocks, not the cover of plants, that create the dominant character. Mountains are often likened to walls, not only as barriers but also by material. There is a mystery to rocks. Solid, they are enclosures that encase and hide. We see only the crust of the earth; only occasionally do those aspects of the earth that we have identified as riches appear on the surface. More often, a trained eye and sensibility seeks surface signs to subsurface riches. On a quest, the prospector carefully follows indicators of geologic, vegetative, and hydrologic patterns, materials, and forms.

For over two hundred years, tales of riches, verified by tantalizing samples of gold, emanated from the Rockies. Spanish explorers sought the mythic Seven Cities of Cíbola and the site of Quivira, all supposedly golden cities. In the 1850s, inspired by the California gold rush, a more deliberate quest was on. In the summer of 1858 two parties—one led by William Green Russell, who had joined with a group from the Cherokee nation, and the other from Lawrence, Kansas—converged on the area of Cherry Creek. There they found "color," the tell-tale glisten of gold in the well-worn rocks of streams coming out of the mountains. They had little real initial success, but enough to catalyze a boom, and the gold rush of 1858-59 began. Silver was soon discovered as miners moved into the mountains west of Denver. Central City, Black Hawk, Georgetown, and Idaho Springs all boomed in turn, as did diggings in and above South Park: Fairplay, Tarryall, Oro City, Buckskin Joe, and Breckenridge. This was only the first of many rushes in Colorado history. The grandest were

## RUSHES

**HARRY H. BUCKWALTER**
Carbonate Hill, Leadville, ca. 1893. *Courtesy of the Colorado Historical Society.*

Leadville's silver boom in 1877 and gold again at Cripple Creek in the 1890s. However, each region would experience its own rush; Aspen, Creede, Silver Plume, the San Juan Mountains and Silverton, Ouray, and Telluride. The ancient allure of gold and silver was not the only attraction. The oil-shale boom of 1915 and again in the 1970s, the Rangely oil fields in the 1940s, and the uranium "fever" of the 1940s and 1950s all promised vast and instant wealth.

Rushes go through cycles. First there is discovery, the exhilaration of a find and the desire to conceal the information. This is inevitably followed by the revelation of the find, exuberant and exaggerated publicity, and a rapid movement of people to the source. The pace quickens with a frenzied assault of the earth, with great riches for a few but disappointments for most. Typically there is a precipitous rise coupled with an equally steep decline once the resource is exhausted. Sometimes, when the initial dust subsides, there is a settling in, with the continued and gradual exploitation of the resource.

A rush concentrates and condenses history. Single-focused and present-oriented, rushes are fast, foster rapid change, and remove things and people from their path. They unite people, resources, and attention. The names given to rushes are revealing. Rushes are likened to stampedes, uncontrollable mass movements of previously placid groups that ignore everything in their path. William Byers, the founder of the *Rocky Mountain News* in 1859, referred to the gold rush as Rocky Mountain yellow fever, a disease. Others likened it to a sacred quest, a gold "crusade" or a hegira to the mountains. "Men are perfectly wild and crazy," wrote a miner from the Gregory diggings in July 1859. All of these emotions and characteristics are expressed in the landscape.

The Kansas City *Journal of Commerce* headline on August 26, 1858, read: "THE NEW ELDORADO!!! GOLD IN KANSAS TERRITORY!! THE PIKE'S MINES! FIRST ARRIVAL OF GOLD DUST AT KANSAS CITY!!!" A week later, the *Kansas Weekly Press* of Elwood was more succinct, with ascending exclamations: "GOLD! GOLD!! GOLD!!! GOLD!!!!" Guides to the gold country appeared almost instantly. Best known was the Oakes and Smith *Guide to the Gold Mines on the South Platte*. It gave data on distance, crossings, and the needs and dangers en route to the gold: water, wood, Indian tribes, grass for cattle and horses, game, and mail stations. Parsons's 1859 guidebook proclaimed the whole Rocky Mountain country to be auriferous. He itemized the components and cost for an outfit for

**WILLIAM HENRY JACKSON**
Upper Creede, early 1880's.
*Courtesy of the Colorado Historical Society.*

six months: $603 and 3,173 pounds, not counting three yoke of oxen and a wagon but including 1,000 pounds of flour. There was advice on rifles, clothing, and the preference for mules over horses. These guides were the state's earliest promotional literature, the grandfather of the chamber of commerce brochure.

The rush affected more than the region of the goldfields. Towns on the Missouri River—Omaha, Council Bluffs, Westport, St. Joseph, Leavenworth—geared up to outfit the miners before their six-hundred-mile passage across the plains. In May 1859, in one week alone, 584 teams left Council Bluffs.

The easier passages followed the blazed trails and rutted routes: the Oregon-California Trail up the Platte River and the Santa Fe Trail up the Arkansas, stopping at Bent's Fort at Big Timbers. Slightly shorter, but more arduous, were the numerous trails that went directly west along the Republican River Trail or along the Smoky Hill Trail, which followed Big Sandy Creek, but these offered little water, difficult terrain, and less grass for grazing animals. By May 1859, one could take the stage line, for $125, along the 689-mile Leavenworth and Pike's Peak Express, from Leavenworth to Denver City. These passages soon became two-way routes. Of an estimated one hundred thousand who went west in 1859 to seek gold, about fifty thousand actually got as far as Denver. Most were farmers with no mining experience, and about twenty-five thousand became a backwash of "go-backs," returning east with only tales.

The prairie roads were white with wagons carrying goods, with many argonauts walking alongside. However, this "frontier" experience was soon abetted by modern technology. By 1863, a telegraph line linked Denver to the nation. During subsequent rushes, most people arrived by rail, which had come to Denver in 1870. Telephone wires were strung over Mosquito Pass to Leadville in 1878.

Newspaper accounts both validated finds and debunked them as Pike's Peak "humbug." One exuberant letter writer said, "Gold is everywhere you stuck your shovel." Those immunized against the "fever" offered sage advice and sober warning. A *New York Tribune* editorial (Oct. 2, 1858) advised, "Better dig gold than stand idle, but better plant corn, shoe horses, build houses, or make fences, than do either of these . . . that there is much gold this side of the Rocky Mountains is scarcely questionable; but you are quite likely to acquire a share of it more easily in almost any other way than by digging it."

**JOHN COLLIER** The Oil Town of Rangely, November 1948. *Courtesy of the Photographic Archives, University of Louisville, Standard Oil of New Jersey Collection.* Collier, and many other Farm Security Administration photographers, went on to work around the world for Standard Oil of New Jersey. From 1943 to 1950 Roy Stryker, former head of the Photographic Division of the FSA, directed Standard Oil of New Jersey's international photographic efforts as part of their ambitious public relations

**WILLIAM HENRY JACKSON**  Gold Mining in Boren's Gulch, La Plata County, ca. 1860.
*Courtesy of the Denver Public Library Western History Department.*

In June 1859 the *Tribune*'s famed editor Horace Greeley, along with Henry Villard of the *Cincinnati Commercial* and Albert D. Richardson of the *Boston Journal*, visited the Gregory diggings. Greeley's enthusiastic report, originally published as an extra of the newly established *Rocky Mountain News*, was widely reprinted in the East but rarely with his warning "not to diffuse a part of our statement without giving substantially the whole," which told of the hardships of crossing to the fields and the need for "capital, experience, energy and endurance" in the gold-mining business. He predicted, accurately, that "thousands are destined to encounter lasting and utter disappointment."

As rapidly as the mountains were washed and dug away, towns were raised, in an oddly additive and subtractive process. The process of town making, typically the slow accretion of settlement through generations, was encapsulated in time-lapse fashion as mining camps evolved within months into towns and cities. Most devolved equally rapidly, their history contained in only a brief period of efflorescence.

In May 1859, John H. Gregory found gold in a gulch along North Clear Creek. The area would soon bear his name. Five weeks later, A. F. Garrison described the Gregory diggings as "past all comprehension." He added, "I have seen at least 3,000 persons, 200 to 300 sluices, and 150 cabins already." In six square miles, he estimated four hundred to six hundred cabins and one thousand to two thousand sluices, and he predicted the numbers would double in another five weeks. The *Missouri Republican* (Sept. 5, 1859) reported on Gregory Gulch: "They naturally hit upon the idea of starting a `town'; a thing easily done in this fast age, and exchanging their canvas abodes for more substantial log cabins.... To conceive the idea, and to commence to carry it out was the work of almost the same moment." These were instant settlements. Webster Anthony, who set up the general store in Oro City, said, "The streets appear as though every one built his cabin in its own place without regard to survey and as a consequence they are very crooked."

The compressed streets of the towns were lined with false-fronted buildings. The false facades often obscured the hills, rocks, and diggings beyond, creating the illusion of a city space in the midst of the mountains. The false fronts were part of these towns' tenaciousness and urban aspirations. One-street, one-story towns had bigger dreams. If there was a collective urban desire, a personal domesticity was equally eked out of the land. As hard

**T. C. MILLER (attr.)**  Hydraulic Mining at Alma, ca. 1881. *Courtesy of the Denver Public Library Western History Department.*

fought as wresting gold or silver from the rocks was the wresting of a settlement from the land, especially the creation of a domestic landscape of home, schools, library, and gentility. The white pickets of fences defined not just a territorial space but a desire and an expectation of behavior, a domain for women and children, for families were the true sign of a settled community.

When the speculator William H. Larimer, whose "speciality was townmaking," arrived at Cherry Creek, two town companies had already been formed, Auraria and St. Charles (which existed only in name). Larimer platted a new town, named it Denver, and merged it with St. Charles. By the fall of 1859, town lots were two hundred to three hundred dollars, up from fifty dollars in the spring. Denver's first structures were tents; the canvas was later recycled as the roofs of the first cabins. The Denver Hotel was built of cottonwood logs with a dirt floor below and a canvas roof above. The construction of walls, roof, and floor symbolized new-found stability and respectability as log cabins gave way to wood frame structures and then brick buildings. Raw nature was cleaned and spruced up. Window holes were cut out as glazing replaced canvas panes, floorboards covered the earth, and paint covered the walls. In December 1859, D. C. Collier wrote, "Our houses are built of logs, often finely hewed, with ground floors and flat roofs, made of poles covered with grass or rushes and earth, with sod, stone and stick chimneys." Just four months later, "We have been assuming more and more the appearance and habits of civilization. . . . The mania for town building is as great as ever."

By November 1859 the alternately dusty or muddy streets of Denver/Auraria had twenty-five general grocery, liquor, and provisions stores, two jewelry stores, two dry-goods stores, five bakeries, ten restaurants, seven hotels, six boarding houses, fifteen saloons, six saloons with billiard tables, four ten-pin alleys, ten doctors, eight real estate offices, six engineers and surveyors, two livery stables, one theater and another under construction, one minstrel opera house, one school, and "2 or 3 churches *talked of.*" Bridges across Cherry Creek and the Platte were being readied. There were also sawmills, liveries, Indian traders, attorneys, a newspaper, a drugstore, barbers, blacksmiths, and gunsmiths.

The camp character of inflated prices, flaunted luxuries, and all-night saloons, gambling, and prostitution coexisted with a rush to civilize. Hotels, schools, churches, lodges, a post

**Photographer Unidentified**   Colorado School of Mines Students on the Reliance Gold Dredge at French Gulch near Breckenridge, 1906. *Courtesy of the Denver Public Library Western History Department.*

office, a library, banks, public baths, and an icehouse were built. Auraria even had a night school for learning French and Spanish. Heritage was installed almost instantly. By 1866 a Pioneers Association had been formed, with membership restricted to those who had come in the rush of 1858-59. Like collective family portraits, commemorative photographs often show civic trappings, celebratory reminders of newly created community pride: banners across streets, bunting around buildings, the July 4th parade, the volunteer fire company on display.

This town-making pattern of rush, speculation, and civilizing was often repeated. In 1879 Leadville had thirty sawmills producing 3.6 million board feet of lumber per week to keep up with the building boom, denuding local hillsides in the process. These were towns at the edge, the margins of civilization, but towns eager to import the best and to be connected to the wider world. Leadville, then Colorado's second-largest city, had 10 saloons, 110 beer gardens, 118 gambling halls, and the Leadville Illuminating Gas Company to light its streets. During the 1895 Crystal Carnival, an Ice Palace, fifty feet high with eight-foot walls, was constructed. A popular tourist attraction, it lasted a full year before melting.

Rapid rise was often followed by even more rapid destruction. Fires were endemic to gold and silver camps. The dense crowding of wooden structures and poor fire-fighting equipment were an invitation to disaster. Georgetown, with four hose companies, never burned, but few other towns were spared. Central City burned in 1874 and was rapidly rebuilt, this time with stone sidewalks and graded streets. Creede was half destroyed by fire in 1892, and Cripple Creek had two fires in five days in 1896.

## DIGGINGS

Colorado is a geological wonderland. A mineral belt 250 miles long stretches from Boulder to Montezuma counties. Here are found gold, silver, lead, zinc, copper, tin, tungsten, and molybdenum. Uranium and vanadium are found in the sedimentary deposits of the Colorado Plateau, and oil shale deposits abound in the state's northwest quarter. In his survey, Hayden said: "At Colorado City there is an area of about ten square miles that contains more material of geological interest than any other area of equal extant in the west. . . . To the geologist Colorado is almost encyclopedic in its character, containing within its borders nearly every variety of geological formation." He added his belief that the Colorado mining districts would be "among the richest the world has ever known."

**HARRY H. BUCKWALTER**  Mining Tramways, Aspen Mountain, February 20, 1893. *Courtesy of the Colorado Historical Society.*

Miners filed claims for the right to extract resources. In the codes of the territory, the claims were defined by type: a gulch (patch) claim was one hundred feet long and went wall to wall in a gulch; a mountain (lode) claim was one hundred by fifty feet; and a creek (placer) claim was for water rights. Later a tunnel claim was added and defined to be two hundred feet along the surface of an outcrop, but in a calculated gamble, the claim was made before digging commenced. There were also timber, ranch, and cabin claims. Under the Federal Mining Law of 1872, the maximum quartz claim (lode) was fifteen hundred by six hundred feet (20.66 acres), measured on the surface. Plat maps of mining claims show numerous bands of these overlapping rectangles. In 1892 Richard Harding Davis wrote that the claims at Creede, with "little heaps of stone and their single stick with their name scrawled on it in pencil, made the mountains look like great burying grounds."

The word *gulch* first appears in American English in 1832, denoting a steeply walled, narrow, mountain ravine, canyon, or gully with water running down its course. Typically inaccessible, the compressed space accentuated the activity between the rock walls. The place name became a verb—one could "gulch" for gold—and in opposite fashion, the word *diggings* came to describe the place. Settlements were built around gulches, and along the gulch were found the apparatus, paraphernalia, and equipment of mining.

Along streambeds, banks, and gulches were placers, natural mineral deposits accumulated through the mechanical action of moving water. Placer mining is a form of surface mining that accelerates and concentrates this natural process. The question is how to screen and separate the "free gold" from gravel, mud, or dirt. Different methods are all variations on the same principle of allowing the heavier gold to settle while water washes away other particles. The most basic method was panning, but the technology rapidly progressed until gulches soon became vast outdoor processing plants filled with rockers (also called cradles), and then long toms, all interconnected by the troughs of innumerable sluices. A *sluice* was either an inclined water channel cut directly into the ground or, more often, a long open wooden box traversing over the surface. Riffles were placed across the bottom of these devices, where the gold, amalgamated with mercury, would collect.

A gulch might initially have provided sufficient water for the process, but with the volume and intensity of activity, demand soon exceeded supply. Gravel and water needed to be

brought together. Gravels were dug, lifted, and hauled from streambeds, banks, and hillsides by progressively more complex devices and methods: shovels, wheelbarrows, tracks, wagons, derricks, buckets, cableways, booming. Ditches and flumes were built to channel water for operations and for the mills that crushed the rock finer and finer until it could be properly smelted.

At Russell Gulch, near Central City, Green Russell, joined by hundreds of miners, built a twelve-mile Consolidated Ditch, bringing water from Clear Creek. It opened on July 4, 1860. The Fair Play Gold Mining Company had two flumes to bring water to its claims, one of which was four thousand feet long by six feet wide and seven feet high. By 1870, Summit County had over one hundred miles of ditches and flumes to supply water for booming. Along the Dolores Canyon, fragments of a flume completed in 1891 still cling to the cliff face, where the flume originally had hung precipitously to five miles of the canyon wall.

Techniques accelerated beyond sluice mining to booming, hydraulicking, and dredging. With booming, a stream was dammed, and water was then rapidly released in a flushing action that removed "more dirt in a single hour than could be formerly excavated in a day," according to an 1872 report. Along the way the ground was stripped of all rocks, trees, and wildlife. With hydraulicking, water was dammed, piped to a site, and then shot at hillsides through nozzles called hydraulic giants. These great water cannons washed away all surface layers until exposing the bedrock. By the late 1860s, surface mining operations were increasingly consolidated and hydraulic works expanded. These methods of greater volume and intensity had a corresponding impact on the landscape as the land was stripped away in an insatiable quest for ore. In 1902 the Keystone Hydraulic Mining Company built a dam across the east fork of the San Miguel River leading to a half-mile, six-by-twelve-foot flume connected to a pipeline and ending at two ten-inch giant nozzles—all capable of washing up to twenty thousand cubic feet of gravel per day.

Dredges, slow-moving factory ships operating in small ponds of their own making, were a later introduction. Methodically, they dug through streambeds thought "played out," for previous placer mining techniques extracted only 60 to 80 percent of the gold. Dredges made possible Breckenridge's third boom (until its ski bonanza). For thirty years, giant electrically

**ROBERT G. ZELLERS**
Tramway and Mill, Shenandoah-Dives Mine, Silverton, 1942–1946. *Courtesy of the Colorado Historical Society.*

powered dredges operated on the Blue River, the Swan River, and French Creek, while the dredge at Fairplay operated until 1952.

Initial discoveries were in the more easily worked sands and gravels of placers and stream terraces. Once these were played out, the adjacent hills were burrowed into. Low-grade ore, too unprofitable at the moment to process, was dumped, leaving it to later technologies to rework these wastes. The methods progressed in an attempt to retrieve it all—to leave no stone unturned and uncrushed. The land is the raw material for the mining process and the stage for the activity. Mountains are the ingredients. Great forces are brought to bear on small areas; for every ingot, a creek or hillside is transformed. It is heavy work, measured in tons of ore, miles of track, and millions of dollars, all to get ounces of metal.

Each aspect of the mining operation has a landscape component: exploration, extraction, production, distribution, and adandonment. The valued deposits are revealed in several ways. In surface mining, the earth's skin is flayed, whereas in subsurface mining, or the lode mine, shafts are dug and miners enter the earth. Minerals emerge from the mouth of these artificial caverns. These small earth orifices only suggest the networks of deep and labyrinthine tunnels for which they are the entry and exit point. Working in subterranean tunnels, miners drill and blast the rock and are followed by muckers, who shovel the material, and by trammers, who push the ore cars.

Little gold is found "free." Most is embedded in ore deposits, which must be milled. At first *arrastras*, Spanish stone grinding mills, were used, but they were not very efficient. Grinders and crushers were successively powered by oxen, water, steam, and electricity. At stamping mills, weights were dropped to crush the ore before smelting. The initial smelting methods were amalgamation techniques. Mercury and the crushed gold or silver were combined, the mixture was heated, and only the mineral remained. Another process was roasting, in which ore was combined with sulphur and then burned off; in the chlorination process, gold chloride was made and then precipitated; and the cyanide process used potassium cyanide, separating the resultant mixture through electrolysis. These processes all required abundant fuel and left ample waste. Wood was used at first, then charcoal. Later a coke industry was born at Trinidad and Cokedale. Originally, smelting operations were

OLA ANFENSON GARRISON
Oil Shale Retort Near DeBeque, ca. 1913–1915. *Courtesy of the Colorado Historical Society.*

**RUSSELL LEE**  Results of Deforestation During the Early Mining Days, San Juan County, September 1940. *Courtesy of the Library of Congress, Farm Security Administration Collection.*

near the mines, but after Nathaniel Hill relocated his smelting operations from Black Hawk to Argo in 1878, regional smelters became the norm.

How did miners bring goods, equipment, ore, and fuel in and out, up and down? The vast quantities of ore, supplies, and men needed to be hoisted and hauled, and for all of these operations, transportation was a constant problem. There were pack trains, toll roads, stagecoaches, and most important, the railroad. Railroads were originally invented in Britain for use in mines, and they were no less essential in Colorado's mining districts, whose mines became subterranean train networks. On the surface, narrow-gage rails of three feet replaced the standard gage of four feet and eight inches. Narrow-gage were easier and cheaper to lay, could be laid in sharper curves through the mountains, and could climb steeper grades. In the 1890s, electric-powered trams were employed, especially on the steep slopes of the San Juan Mountains at Silverton, Ouray, and Telluride, to move buckets of ore up and down mountainsides. All of these elements were linked: mines, water, mills, and smelting operations. The railroad fused the operations within Colorado and ultimately connected them to a national network.

The contrast between the subterranean, troglodytic world of the mine and the surface world was striking. Mines were dark, wet, cold, black, smoky, fetid, cramped corridors. Light came from oil lamps, later from naphtha or gas, and finally from battery-powered lights. Equally striking was the difference between the mines and the interiors of the town's bars, dance halls, hotels, restaurants, theaters, and opera houses. Mines were damp, with the unseen danger of mine gases and the horrors of collapse and cave-in, whereas these pleasure places were opulent, gay, colorful, and sparkling. Compared with the mines or the squalid cabins, these places glistened like gold.

The contrast in mining towns was equally true for front and back. The false-fronted streets put on their best face while the rear yards were crammed with piles of storage, garbage, and outhouses draining directly into streams. In April 1879, five thousand wagonloads of filth were removed from Leadville's streets and alleys in an early clean-up campaign.

Photographs can show us the look, but it is difficult for us to sense the mining landscape—its noises, smells, heat, and vibrations. These were the sounds of power, and they continued night and day. Below were miners who drilled and blasted with black blasting powder and,

**RUSSELL LEE**  Tailings Pit of the Gold Mill, Telluride, September 1940. *Courtesy of the Library of Congress, Farm Security Administration Collection.*

after 1864, dynamite. Engine houses, pumps, and compressors forced air into shafts for ventilation, pumped water and noxious gases out, and powered hoists and trams. The braying of mules, used in the mines until the 1950s, the buzz of sawmills, the clang and fire of ironworks, and the rushing of waters all contributed to the noise. The mining landscape is one of movement, loud and visible processes and activity, mud and smoke. Matter is present in all its states, as solids, liquids, and gases change back and forth. The boom-and-bust cycle had an aural parallel, as the land went, in the span of a few years, from the sounds of nature to the bustle of activity to the hush of inactivity.

No longer do we sense the most common, but ephemeral, landscape experiences: the whistle of trains, the clang of fire bells, the billows of engine smoke, the watchfulness for cinders, the odors of decay. The plumes of the household hearth have been replaced by collective, ubiquitous, insidious smog. Our most common contemporary sound is that of traffic, which often acts as a form of mechanical Muzak masking the natural sounds of the land.

## TAILINGS

Diggings became "worked out," "played out," "exhausted." What remains are the tailings, the waste piles of the mine. The process is fast, the effects are lingering, and the consequences are rarely anticipated. By August 1859, the *Missouri Republican* reported that the Gregory diggings, where gold had been discovered in May, were played out. "Dozens of idle sluices and slides for transporting 'pay dirt' down the mountain side to the bank of the brook; abandoned cabins, unworked shafts, waterless ditches, could be seen in every direction. The valley was no longer distinguished for want of elbow-room among its sojourners. Not the twentieth part of the dense mass, which moved up or down its narrow bottom at the time of my last visit, remained. . . . How did it happen that the buoyancy, activity, enterprise and success which characterized 'June and July' made room for depression, dullness, relaxation and failure." The Gregory diggings had been "the supposed embryo of a future city," but most of these embryonic cities were stillborn.

As early as 1867, O. J. Hollister in *The Mines of Colorado* could write: "Nothing indeed can be more deceiving or more ephemeral than the feverish prosperity of a placer mining country. California Gulch, which six years ago was infested by 5,000 to 6,000 people, is now

**DREX BROOKS**  Sweet Medicine, Sand Creek Massacre Site, 1987. *Courtesy of the artist.*

"In going over the battleground the next day, I did not see a body of man, woman, or child but was scalped; and in many instances their bodies were mutilated in the most horrible manner, men, women, and children—privates cut out, etc. . . . According to the best of my knowledge and belief, these atrocities that were committed were with the knowledge of J. M Chivington, and I do not know of his taking any measures to prevent them." James D. Cannan, First Lieutenant from: *House Report, Massacre of Cheyenne Indians,* January 10, 1865.

almost deserted. The relics of former life and business, old boots and clothes, cooking utensils, rude house furniture, tin cans, gold plans, worn-out shovels and picks, and the remains of toms, half-buried sluices and riffle boxes, dirt-roofed log cabins tumbling dow, and the country turned inside out and disguised with rubbish of every description was most disagreeably abundant and suggestive."

Each mining extraction and processing method creates its own waste. Until recently legislated reclamation efforts, few cleaned up after themselves. The waste was profligate, and the whole landscape was made to serve the imperative of mineral extraction. The area is a "country turned inside out"—streams ripped open and diverted, vast open pits, carved mountains, and great slipping mounds of tailings beside crushers and mills. Hillsides are denuded and eroding, for the mines were rapacious users of timber resources as fuel and building material and especially for railroad construction. Even where no minerals were present, the hills were "mined" for their resources. There are also the more insidious wastes: polluted soil and water and the poisonous chemical wastes of mercury, sulfuric acid, and cyanide from smelting operations.

Mining waste was typically just deposited or left in the landscape; gulches are littered with rusting equipment, rotting timbers, collapsed buildings, and unmarked shafts. Tailings are the detritus of mining, but also of construction of houses and highways. Each leaves a "tailing" in the landscape, leftovers from digging out the earth, building a foundation, or cutting through rock. Roads and railroads cut and slice the landscape; housing construction carves the land. Some of these effects are "healed" by vegetation, but in Colorado's arid landscapes and beneath layers of snow, the progress is slow. Therefore the human imprint, the record of working or extracting from the land, remains visible long after the cessation of activity. Century-old remnants persist as a relict landscape. The Creede mining district remains as a dramatic exposition of mining, of its ruggedness and confrontation with the landscape of rock. Creede's principal street leads up the gulch and steeply ascends parallel to the creek, now littered with machinery, timber, and tailings. Built within the confines of a narrow canyon, the mine's operations, supported by great bulwarks, recall the scale, mystery, and horror of Giovanni Piranesi's *Prisons*. In the Cuchara Valley, still mined for

**KENNETH HELPHAND**  Ludlow Monument, 1989.
*Courtesy of the author.*

"In memory of the men, women and children who lost their lives in freedom's cause at Ludlow, Colorado, April 20, 1914. Erected by the United Mine Workers of America."

coal, at Cokedale are the ruins of coke ovens and mountainous, giant black tips, flocked in summer with yellow flowers.

The state has vast resources of fossil fuels, oil shale, and coal underlying the Western Plateau and portions of the Piedmont. They form an invisible layer, which, like the gold and silver of the nineteenth century, certainly will influence the century to come. The problems of extraction are many. Typically, the material is removed through surface mining, where the overburden covering the vein is excavated. It is like carving away skin and muscle to get to the organs of an animal. Even if the land is reclaimed, this is a form of earth taxidermy. Residents of Meeker refer to May 2, 1982, as "Black Sunday," when Exxon closed the Colony Project, signaling the end of the most recent oil shale boom. Meeker had doubled in population, but now oil shale operations stand closed, awaiting their inevitable development.

The 1940 "uranium fever" swept western Colorado, encouraged by the Atomic Energy Commission, headquartered in Grand Junction. Prospectors, armed with Geiger counters, came by airplane and jeep. Processing plants were constructed at Rifle, Uravan, Slick Rock, Maybell, Durango, Naturita, Gunnison, Grand Junction, and Cañon City. The boom was over by the late 1960s, but the effects linger, for these tailings are the most frightening. The radioactive wastes from the mining of uranium and vanadium do more than pollute a landscape. The transmission of radiation is invisible. It remains an insidious contaminant for generations, and the effects are the debilitation of disease and the randomness of mutation and defect.

The ill effects of mining were not all environmental. There has been a social and cultural cost as well. There are many kinds of "tailings." The frontier democracy of cash and opportunity was not universal. Chinese laborers were brought into the mines as the processes became more industrialized. At Fairplay, tents and cabins were set up for Chinese laborers. Across the river or at the edge of town were boomtown ghettos, named "Chinaman's Gulch" by the miners. Meanwhile, the region's ancient inhabitants, the Indians, remained in their villages, but their immemorial claims and treaty rights were tragically betrayed. No accommodations were made to the tribes, as the westward movement of miners accelerated the fate of the native inhabitants. The 1880s saw a concerted effort at "Ute removal," forcibly relocating the inhabitants of the West Slope and eliminating them as an obstacle to mining and white settlement of the region.

Two of the state's most poignant places are landscapes of memorial and memory. Sites of horrible events, they speak of human degradation and waste. The land is not poisoned, but the very ground remains consecrated by the dead. The Sand Creek Massacre site is marked only by a plaque in the midst of silent grasslands. It is left to your imagination to hear the echoes from the barbaric November 1864 massacre of two hundred to four hundred (the number is debated) Cheyennes and Arapahos, mostly women and children. The nearest town is Chivington, named for colonel who led his forces into "battle."

At Ludlow, within sight of the freeway, stands the Ludlow Monument to the victims of the "Battle of Ludlow" during the 1913-14 coal strike. In a confrontation between striking miners and the state militia, the miners' tent colony was burned. Five miners, four militia members, and two children died, along with two women and eleven children who suffocated in a cellar of a burning tent. A memorial sculpture has been erected at the actual site of the tragedy, showing a family grouping of a miner and a mother clutching her child. A few feet away is a metal cellar lid. Outside the small fenced enclosure is a register largely filled with the signatures of members of the United Mine Workers of America, who invariably add their UMW local number. This is a pilgrimage site, a mine worker's shrine.

Charles Henderson concluded his classic 1926 work *Mining in Colorado* by critiquing those who said Colorado's surface had only been scratched. Not so, wrote Henderson. "The surface has been well scratched and even intensively perforated with holes ranging from 10 to 3,000 feet in depth and with tunnels as much as 5 miles in length. Much of this `scratching' was misdirected" by those ignorant of minerals and mining. Our knowledge of the land—its elements, dynamics, and processes—is certainly not complete, but we now know the impacts, both short- and long-term, of these activities. That knowledge must be part of the environmental, as well as economic, accounting as the continued value of mining is assayed.

**KENNETH HELPAND**  Near the Headwaters of the Rio Grande, Mineral County, 1989. *Courtesy of the author.*

# WATER

## *"Adam's Ale."*

THE HEADWATERS of the Colorado, Rio Grande, Arkansas, and Missouri rivers lie in the Colorado Rockies. Dropping thousands of feet, they cascade down mountainsides, slither across mountain meadows, roam the floors of valleys, and slice through gorges. White foam and rapids gather materials and become the brown liquid of the rivers of the plains and the Colorado Plateau. In following the course of rivers, one experiences the state's true diversity. Dr. Edwin James, of the Long Expedition, was along the Arkansas River, east of Pueblo, when he observed: "At the base of the mountains, the water was transparent and pure, but soon after entering the plains it becomes turbid and brackish.... The boulders, pebbles and gravel so abundant near the base of the mountain, had been growing gradually less frequent and diminishing in size, till they now almost entirely disappeared, their place supplied by fine sand intermixed with clay." Samuel Bowles said that in the gorges of the mountains, the rivers "`canyon,' as by making a verb out of the Spanish noun, the people of the country describe the streams as performing the feat of such rock passages." The river valleys are distinctive landscapes defining territories and local cultures of ranching, agriculture, recreation, or wilderness, from the grand but dramatically different Gunnison, Dolores, Yampa, Green, and Arkansas to the smaller Huerfano, Saguache Creek, Animas, and Blue river valleys.

Testimony to the quality of Colorado water predates modern advertising hyperbole. In Sam Hawken's August 8, 1859, letter from Denver, he described the water as "Adam's Ale," flowing out of the mountains as clear as crystal. His image is striking and true, for the waters are both paradisiacal and intoxicating. Elevation is again the key, as the mountains catch the clouds moving from the west. Alpine areas receive thirty to sixty inches of precipitation,

## PATTERNS AND PERCEPTIONS

mostly in the form of snow, but most of the state is semiarid. Cities along the Front Range receive around sixteen inches annually, the plains a bit more; a few areas, the San Luis Valley and some western valleys, receive only seven inches per year. The driest areas abound with intermittent streams. There are few natural lakes, yet there are vast subsurface water resources, particularly in the east.

The historian Donald Worster has applied the term *hydraulic civilization* to the arid zones of the American West. The term was originally coined by Karl Witfogel to describe the ancient cradles of civilization. He hypothesized that the control, regulation, and distribution of water was the essential determinate of social and political organization. Worster deems the West to be a modern hydraulic society whose social order is "based on the intensive, large-scale manipulation of water and its products in an arid setting." The Colorado author Thomas Hornsby Ferril has observed, "Here is a land where life is written in water."

Over 13.6 billion gallons of water are used daily in Colorado. Surface water supplies 82 percent, the remainder coming from groundwater. Irrigation uses 91 percent. Just over 5 percent is for the public water supply while industry uses half that. In a land with scare rainfall, the signs of water—its capture, storage, and distribution—are everywhere, if one chooses to look. Water is captured from melting snows and mesa tops, pumped from subsurface aquifers, diverted from rivers, and stored in towers and reservoirs. It is channeled, funneled, pumped, siphoned, and sprayed, directed and redirected for purposes practical and pleasurable. In most towns it is the water tower, not the church steeple, that soars highest. One significant state measure is the acre-foot, one acre of land covered with one foot of water—325,851 gallons.

Engineering and design have tried to overcome the basic constraints of water scarcity and location to enable large numbers of people to dwell in this landscape. For Walter Prescott Webb, the author of *The Great Plains*, the West was an "oasis civilization," whereas Marc Reisner, in *Cadillac Desert*, is harsher in his characterization, describing habitation in the arid West as a "beachhead," where "everything depends on the manipulation of water—in capturing it behind dams, storing it, and rerouting it in a concrete river." The natural order of Colorado land is controlled and mitigated, as water has been given a revised shape and

**Photographer Unidentified**
Man Opening Primitive Wooden Valve on Irrigation Ditch Near Grand Junction. *Courtesy of the Colorado Historical Society, Rio Grande Collection.*

**JOSEPH BEVIER STURTEVANT**   Elaborate Fencing and Landscaping Along Farmer's Irrigation Canal at Corner of 6th and Spruce Streets, Boulder, ca. 1900. *Courtesy of Carnegie Branch Library for Local History, Boulder Historical Society Collection.*

pattern. River channels and courses have been stabilized and their flow regulated. "Lakes," actually storage reservoirs, exist where streams flowed. Water is diverted across watersheds; dry lands are irrigated; and groundwater is withdrawn, with little replenished.

Use and history determine our relationship to water. Even moving and opaque water acts as a mirror, reflecting our ideas about what we see. It is a truism that we all perceive the world with different "eyes" as we all "see" the world and create our own mental images. When we share images, communication is easy; when images are different, it may be difficult. Water is the primary component and symbol of life; it is nature unfettered, determining and following its own cycle and course. For farmers and ranchers, water sustains crops and animals; for fishermen, it is a recreation site; for engineers, it is a source of power. Water is essential for habitation. However, for most urbanites, it arrives via a tap and has a brief cycle—faucet to basin to drain—a miniscule segment in a grand hydrologic scheme. Water is to swim in, drink, and irrigate crops; it is a sewer, a road, a place to "run"; it is a hazard. It may be all of these simultaneously. To understand water, one needs to know not only its geography, hydrology, and history but also the diverse water "eyes," the images of others, for perhaps in no other aspect of the landscape are perceptions more diverse. For many, water is a "resource" to be accounted for, ordered, and used, above all not to be "wasted." Felix Sparks, the former head of the Colorado Water Conservation Board, rhetorically once asked: "What are you going to do with all that water? Are you just going to leave it in the ground?" For Donald Worster, the irrigation ditch symbolizes a "sharply alienating, intensely managerial relationship with nature"—"simplified, abstracted water." Colorado water is pragmatic and pleasurable, a resource, commodity, and property, but it also provides the setting for camping by a mountain lake, kayaking through some rapids, and skipping stones across the Platte.

On the wall of the Gateway Cafe, almost in Utah, is a precipitation chart, an accounting and sign of hydraulic civilization at the most local level. In the summer of 1989 it read:

*Annual Precipitation*

| | | | |
|---|---|---|---|
| 1980 | 20.01 | 1985 | 15.43 |
| 1981 | 16.22 | 1986 | 12.85 |
| 1982 | 15.14 | 1987 | 17.07 |
| 1983 | 17.73 | 1988 | 11.02 |
| 1984 | 15.40 | 1989 | — |

A second table provided monthly averages over ten years, and another gave the monthly figures for the current year. The ten-year average was 15.16 inches; by mid-August 1989, the total was 5.94 inches.

## DITCHES

The historical lineage of Colorado irrigation is long. Farmers who came north to southeastern Colorado brought Mexican irrigation techniques developed by the Spanish, who had learned them centuries before from Arab conquerors. Settlers from eastern, temperate states, accustomed to regular and reliable precipitation, had to learn the ways of a climate of great variation and unpredictability. In early state histories, the question of who were "the first irrigators" became something of a matter of state (and ethnic and racial) pride. The Mormon experience in nearby Utah was denigrated, as were Hispanic and Native American contributions. In reality, each group had to learn its own landscape lessons; Anglo settlers relearned Spanish wisdom, but neither Anglos nor Hispanics learned from the land's native inhabitants.

The systems all have elements in common, from that of the Anasazi to the system of the farmers of the South Platte Valley to the great circles of pivot-irrigation systems on the plains. Water must be captured, diverted, channeled, stored, distributed, managed, and monitored. Technique, scale, pattern, and intensity vary from simple diversions to systemic networks covering hundreds of square miles. All ultimately deliver water to fields, furrows, and individual plants. Historically, the first stage was diversion ditches, which directed streams and creeks to lowland fields. Ditches were then extended into higher terrain, and subsequently storage basins and reservoirs were built to even the flow and to protect against

**Photographer Unidentified**
Irrigation Flume in the Grand Valley Near Grand Junction, ca. 1884. *Courtesy of the Colorado Historical Society, Rio Grande Collection.*

**Photographer Unidentified**  Skinny Dipping, South Canal of the Gunnison, Division 2, Largest Drop, July 9, 1908. *Courtesy of the Colorado Historical Society, Rio Grande Collection.*

drought. Ultimately, natural watershed boundaries were breached, and water was diverted to new lands, creating a revised hydrologic map of the region.

The Pueblos and their Anasazi antecedents were the area's earliest irrigators. Mesa Verde, the "Green Table," has been called an oasis in the sky, made possible by the precipitation at its higher elevation, which starkly contrasted with the surrounding arid lowlands. Water was managed of necessity—on the mesa are found the ruins of collection channels, ditches, and a reservoir system including Mummy Lake, ninety feet across and twelve feet deep. At Wetherill Mesa are over one thousand rock check-dams built to stem erosion from the farmed mesa top, as well as small agricultural terraces in gullies. Domestic water was collected in seeps, the springs along the cliff faces that supported dwellings.

Anglo and Hispanic settlers and irrigators arrived almost simultaneously. At San Luis is the San Luis People's Ditch, which diverted water from the Culebra River and has been in continuous operation since April 10, 1852. Its claim as the first appropriated water in the state (Decree No. 1) was confirmed in 1889, when nine hundred acres were still farmed by descendants of the original settlers. Nearby are Decree No. 2, the San Pedro Ditch near Chama, and No. 3, the Acequia Madre on the Costilla River. The Hatcher Ditch (also known as the Lewelling Ditch and the Lewelling-McCormick Consolidated Ditch) was dug in October 1846; its use stopped in 1847, only to be revitalized in 1865. It has been in continuous use since that date, diverting water from the Purgatoire at Hole-in-the-Prairie, twenty miles from Trinidad. Hatcher, a foreman for Bent, St. Vrain and Company at Bent's Fort, originally planned to use the ditch to grow hay for ox teams.

In 1848 George Ruxton observed, "Irrigation is indispensable over the whole of the region, rain seldom falling in the spring and summer, which is one the greatest drawbacks to the settlement of this country, the labor of irrigation being very great." Edward Bliss, in an 1861 official report on the "Soil, Climate, and Resources of the Territory of Colorado," reported, "Nature withholds her rains from the plains of Colorado and admonishes her people that they must strive to overcome this deficiency by artificial purposes." He added that for mountain waters, "the sturdy farmer has but to check the dashing torrent in its course, and divert a portion of its grateful element, through artificial channels, across his thirsty and arid fields." By 1864 the *Rocky Mountain News* was urging a plan whereby a portion of the public domain would be granted to those who constructed irrigation canals. A decade later President Ulysses Grant recommended a canal for the purpose of irrigating from the eastern slope of the Rockies to the Missouri River. Cyrus Thomas, writing in the 1873 Hayden report on the "Agriculture of Colorado," noted, "Water is the great desideratum of the agricultural development of this country, and the method of its distribution we shall find is the true key to the agricultural system of the Territory."

The progress of irrigation followed a set of topographic imperatives, beginning with crude bottomland ditches used to water hay lands. Canals were then extended to benchlands and tablelands, followed by longer and longer canals beginning at higher elevations. The hydraulics of mining and agriculture occasionally worked together, supplying water for both placer farming and farming. As the scale and scope of systems expanded, cooperative and corporate ventures became inevitable.

The most dramatic venture was the transformation of the Piedmont and the pioneering efforts of the colony towns north of Denver. In 1869 Horace Greeley sent Nathan Meeker, the agriculture editor of the *New York Tribune*, to study Mormon irrigation practices in Utah. Meeker never got that far, but he was most impressed by Colorado. In 1871, shortly after Denver was linked to the national rail network via Cheyenne to the north and Kansas to the east, six hundred members, mostly heads of families, pooled their resources, purchased along the Cache La Poudre River, and founded Union Colony. One of their first acts was the construction of a forty-five-mile perimeter fence defining their domain. The town was named Greeley.

The inexperienced irrigators dug their first canals and learned their lessons by trial and error. Slopes were too great, water was lost through seepage, capacities proved inadequate, supplies were insufficient, and the ultimate cost for ditches was twenty times that originally budgeted. However, canals were completed: the longest was an impressive thirty-six miles long, twenty-two feet wide, and five feet deep. A lateral canal coursed through the city streets, providing water for household use and irrigation for trees and flower beds along the way.

Greeley's canals were Colorado's first large community cooperative canals and the first to irrigate benchlands. They demonstrated the ability of corporate activity to develop water resources. A utopian idealist, influenced by the ideas of Charles Fourier, Meeker found in the practice of irrigation an operational equivalent for his communal ideals. In 1876 he reported on Union Colony, "It was discovered that selfishness and individualism had no place in dealing with the element that is the lifeblood of agriculture." Water came with the purchase of rights by colony members, with an annual charge added for superintendence and repair, about twenty-five cents per acre. Thousands of acres were placed "under ditch," the term for lands so situated that water could be obtained from an existing ditch. Meeker had imagined an intensive agriculture, but it became extensive. By the mid 1920s, three hundred thousand acres were under ditch in Weld County alone.

Other colonies and canals followed, and during the 1870s thousands of acres were put under ditch along the tributaries of the South Platte. In 1872-73, the Fort Collins Agricultural Society was organized by R. A. Cameron of Union Colony; Longmont was founded by the Chicago-Colorado Colony; the Tennessee Colony settled near what is now Orchard; and the St. Louis Western Colony began five miles west of Union Colony. Outside investors, mostly English, joined the cooperative colonies in developing the land. The Larimer and Weld Canal, the second-largest in the state and owned by Colorado Mortgage and Investment Ltd. of London, by 1881 irrigated sixty thousand acres between Greeley and Fort Collins. The English Colorado Mortgage and Investment Company built the North Poudre, Bessemer, Fort Lyon, Bob Creek, and Otero canals. Farther south, the Denver High Line Canal was completed in 1882, moving water forty-four miles from the mouth of Platte Canyon to Cherry Creek. Ditches crisscrossed Thompson Valley, St. Vrain Valley, and Fountain Valley below Colorado Springs. By 1882, the irriguous Poudre Valley was called "one vast network of canals."

Tragedies for one group were opportunities for another. In the western part of the state, the August 1881 expulsion of the Utes signaled the September opening of lands for white settlement. The Grand Valley irrigation projects near Grand Junction began almost immediately. Three ditches—Grand Valley Ditch (12 miles), Pioneer Ditch (6.75 miles), and Pacific Slope Ditch (9.5 miles)—all took water from the Grand River, officially renamed the Colorado in 1921. Running parallel to the river, they were built hastily and poorly and often collapsed; their capacity was low, and they even flooded Grand Junction. The new

Grand River Ditch, completed in 1884, consolidated these projects, remedied the errors in design, and brought all low-lying areas under ditch. Designed by Matt Arch, the ditch was a formidable thirty-five feet across at the base, fifty feet at the top, and five feet deep.

Photographs hint at the substantial human efforts necessary to create these systems. Men worked with picks and shovels, and horses pulled Fresnos, large iron scraping shovels. Areas were dynamited as needed, while carpenters built headgates and flumes across low spots. The mainline headgate was located by Palisade, with the ditch extending twenty-four miles and ending west of Fruita. The most dramatic component of the system was John Wellington's huge waterwheel, constructed to lift water from the ditch fifty feet to his orchards. Originally, the irrigation structures—headgates, flumes, and weirs—were all made of wood, but after the 1889 flood they were replaced by steel and stonework. Over a century later, the Grand Valley Canal system has one hundred miles of ditches delivering water to forty-five thousand acres.

Ditches were used not only in practical ways. Outside Shale City, the fictionalized Grand Junction of *Johnny Got His Gun*, Dalton Trumbo, who grew up there, said: "In the summer they went out to the big ditch north of town and stripped off their clothes and lay around on its banks and talked. The water would be warm from the summer air and heat would be rising off the brown-grey land like steam. They would swim for a while then they would go back on the bank and sit around all naked and tan and talk."

In 1889, the U.S. Census first reported irrigation data. Colorado had 890,775 acres irrigated, second only to California, and from 1899 to 1919, Colorado would lead the nation. Professor L. G. Carpenter of the State Agricultural College reported that almost 3 million acres, almost 4,600 square miles, were under ditch, with a third under cultivation. He took pleasure in placing Colorado in the company of other nations practicing irrigation much longer than the few decades of modern Colorado history, noting that in all of Egypt, only 7,000 square miles were irrigated.

Elwood Mead, an irrigation expert, enthusiast, and publicist, estimated that 4 to 5 million acres of the state could be irrigated. Mead's estimates proved accurate. By the 1920s, 4.7 million acres were irrigated by 28,000 main canals and laterals supported by 1,000 storage reservoirs and dams. Demonstrating an early conservation awareness, Mead urged the

**GEORGE BEAM** Apple Orchard, ca. 1890. *Courtesy of the Colorado Historical Society.* Caption: "Colorado National Apple Exposition, 2nd Prize for Photograph of Irrigation Scene."

protection of timber resources as "natural" reservoirs for watershed lands. It is the "duty of the state to foster its [water] economical use and to restrain, and if possible prevent, everything which encourages wasteful and pernicious habits. The way to accomplish this is to adopt a system of delivering water that will make it to the interest of the individual irrigator to practice economy."

Surface irrigation is accomplished through flooding or furrowing. The choice of method depends on the soil, the topography, the head (i.e., volume) of water, and the crop. The "duty of water," the quantity of water needed to mature crops, is 1.8 feet in the San Luis Basin and 2.2 feet in north-central Colorado. Level land is best for even irrigation; therefore the land is graded and scraped to reduce high points. Contemporary irrigators accomplish this by laser leveling. Ditches are then laid out, with simple ditches plowed or furrowed in fields using a remarkable set of agricultural implements invented to accomplish these tasks. Small dams and weirs (of canvas, wood, or metal) direct and regulate flow. Water losses diminish the efficiency of the distribution system, so ditches are puddled (soaked with water, which settles and minimizes seepage), veneered (sealed with a lining of silt or clay), or made impervious by the construction of wood, concrete, or metal ditches, pipes, and flumes. Each water conveyance system has a headworks, a diversion structure or storage dam leading to a main canal or pipe system and then to progressively smaller canals or laterals with gates to regulate flows. Throughout the irrigation landscape are monitoring devices—gauges to monitor flow, volume, and levels. At headgates, feet are marked off like doorway jambs that measure childhood growth. The ditch rider's task is to maintain the system, keeping channels clear, flows smooth and steady, and allocations as regulated. Water in the Grand Valley, en route to the sea, now finds its way flowing across fruit orchards. In Weld County, the distribution system extends down to its smallest unit, with each furrow feeding off individual siphons taking water from troughs lined along the edge of fields. The furrows recall the orderly lineup of cattle at a feedlot, in what is the nation's second most productive agricultural county.

Water is a political as well as an agricultural issue. State irrigation conventions in 1873 and 1878 led to the passage of irrigation acts and the creation of state water divisions supervised by a water commissioner who could rule and adjudicate. Colorado was the first state to have such a structure, along with the powerful position of the office of state engineer.

O. T. DAVIS  Some Buena Vista Oats, Result of Irrigation on B. McCabe's Ranch, ca. 1898. *Courtesy of the Colorado Historical Society.*

The 1902 Newlands Act creating the Bureau of Reclamation put the federal government in the role of constructing and maintaining irrigation works for water storage, diversion, and transmission in sixteen western states. A Reclamation Fund from public land sales financed the agency's projects. *Reclamation*, in the vocabulary of the Progressives of the day, referred to the reclamation of arid lands for settlement. Ironically, in a region that extolled the virtues of individualism, it took a massive federal presence and vast cooperative engineering to "reclaim" the land.

One of the bureau's first projects was the Uncompaghre Irrigation Project. By this time, the Indians had been removed, the Denver and Rio Grande Railroad had arrived in Montrose in 1882, and extensive ditch digging had taken place. When the Reclamation Service began its work in 1903, private enterprise had already constructed 110 canals and laterals totaling almost five hundred miles. The government acquired those and created a unified irrigation system in the valley. A change in materials was symbolic of the government's efforts: wooden headworks were stabilized and replaced with concrete, and wooden flumes were exchanged for steel ones. The first two years of the bureau's work were spent studying, surveying, mapping, researching, planning, and most important, selecting the canal and tunnel line to bring Gunnison River water to Uncompaghre Valley. The Gunnison Tunnel cut through the wall of the Black Canyon of the Gunnison and beneath Vernal Mesa. Almost six miles long and ten in diameter, when opened by President William Howard Taft in 1909 it was the longest irrigation tunnel in the world. The Gunnison was dammed to divert water into the tunnel's entry, and at the exit, the South Canal took the water eleven and a half miles to the Uncompaghre River to irrigate seventy thousand acres of Delta, Montrose, and Ouray counties.

Early Reclamation Service engineers had to survey drainage basins, locate lands, classify soils, study hydrology, estimate floods, size spillways, and design and locate dams, canals, tunnels, and bridges. In the Grand Valley Project, the Reclamation Service augmented the enterprises that had begun in the valley thirty years earlier. The Grand (Colorado) was dammed eight miles northeast of Palisade. The valley itself had 118,000 irrigable acres, but waters needed to be raised one hundred feet, so a gravity canal was built, which also powered electrical pumping plants to lift the waters. George Wharton James, in his *Reclaiming the*

**GEORGE BEAM** Workers at the West Portal of Gunnison Tunnel, ca. 1909. *Courtesy of the Colorado Historical Society.*

*Arid West: The Story of the United States Reclamation Service,* typified the dominant viewpoint of the time when he said that there was an "unlimited volume of water going to waste annually down the Grand River."

Colorado's surface water resources are substantial; however, in recent decades subsurface water resources (groundwaters) have been the focus of attention. Water has always been hoisted or pumped from wells. Windmills pumped water, but the invention of the diesel-driven centrifugal pump, capable of delivering eight hundred gallons per minute to the surface, made possible the massive development of underground water resources. This was coupled with the invention of the circular pivot irrigation systems now so prevalent on the plains and wherever subsurface aquifers can be tapped. Frank Zybach, a tenant farmer from Nebraska who farmed near Strasburg, Colorado, was granted a patent in 1952 for his center-pivot irrigation machine, which he called a "self-propelled irrigation apparatus." Employing lightweight aluminum pipe and rotating sprinkler heads, these devices now slowly arc across thousands of acres previously classified as nonagricultural land or marginal rangeland. Striking in their patterns from the air, each of these automated great circles, inscribed within square quarter sections, covers 133 acres. Colorado and Nebraska have led the nation in the use of these irrigation machines since the 1960s.

In 1929 there were only eight hundred irrigation pumps in the entire state. Thirty years later, there were eighty-four hundred. Until this subsurface boom, there had been little need for regulation, but the first groundwater law was passed in 1957 and later augmented by the 1965 Groundwater Management Act. In 1985, Colorado used 2.34 billion gallons of groundwater daily, primarily for irrigation; 13.9 million acre-feet of water are annually sprayed and flooded over 3.4 million acres of Colorado fields, with 20 percent coming from groundwater supplies. Almost five thousand new wells are drilled annually, and water, a finite resource, is being mined, not to be replenished in the life of current generations. There is a critical level of groundwater overdraft in most of eastern Colorado above the diminishing Ogallala aquifer. The phenomenon has been described as "underground desertification." The water boom will inevitably bust at current levels of use, and the green circles will once again revert to rangeland.

## DIVERSION

The supply is to the west, the demand to the east. It is a problem in location and distribution, haves and have-nots, with the results changing the fundamental divisions of wet and dry. The transformation of the "natural" landscape order is seen no more dramatically than in the change of river courses. Transmountain water-diversion projects divert water from one side of the continent to the other. Raindrops and snowflakes fated for the Pacific find their way to the Atlantic; instead of a journey through the Grand Canyon, they slowly float the Mississippi.

Diversion projects have created a new map, changing the "natural" catchment boundaries—defined by ridgelines, slopes, watercourses, and valleys—to watershed lines that leap over mountains. Waters taken from one watershed to another are known as "foreign waters." The natural power of water creates and conforms to the topography of the land. In diversions across the continental divide, the imperatives of topography are overcome, giving new shape to Colorado water. No longer permitted to follow its innate imperative of seeking its own path, the waterway is predetermined, its will co-opted as it is channeled, squeezed, smoothed, contained, siphoned, controlled, and pumped.

In the semiarid landscape, this new course of water both created and reinforced the desired pattern of agriculture and settlement. The irrigation boom and canal building of the 1870s and 1880s along the Eastern Slope increased demand for even more water resources, and farmers eyed the abundant waters of the Western Slope. The first diversions were modest in scope. In 1880, water was diverted from the Eagle to the Arkansas River for mining. In 1882, Cameron Pass Ditch took water from the South Platte to the Cache la Poudre watershed, and the 1892 Skyline Ditch first diverted the waters of the Laramie River to Poudre Valley. In 1888, water was brought from the Dolores River to the Montezuma River valley through a mile-long tunnel. These early ditch diversions led to transmountain diversion tunnels, with reservoirs on the Eastern Slope in the South Platte, Arkansas, and Rio Grande valleys in the nineties, but the amounts were all modest.

The demands were not only agricultural. The burgeoning population of Denver needed water. In 1905, the Denver Union Water Company completed Cheesman Dam on the South Platte, then the world's highest gravity-arch masonry dam and the first on-stream storage dam for municipal water in the West. Cheesman hired Charles Harrison to design

a new dam after a previous dam was destroyed while still under construction in 1900. Harrison gave it an innovative arch shape for increased strength and safety. It was 176 feet thick at the base and only 18 feet at the top, 221 feet high, and 1,100 feet long. The dam was built of natural granite blocks quarried on the site, each weighing up to six tons.

Private initiative was superseded by public bodies in 1918, when the Denver Board of Water Commissioners acquired the Denver Union Water Company, which supplied the city's water. In the early twenties, Denver began to acquire water rights along the Fraser, Williams Fork, and Blue rivers. When the Moffat Tunnel, built for railroad traffic, was completed in 1936, the tem-and-a-half-foot "pioneer bore" became an aqueduct, taking Fraser River water to South Boulder Creek for delivery, though a system of reservoirs, conduits, and canals. Many additional diversions have subsequently been completed. The Blue River diversion stores water in the Dillon Reservoir (1963) and then takes it, via the Harold D. Roberts Tunnel, 23.3 miles from Dillon to Grant.

It is not surprising that the key symbolic site of James Michener's epic Colorado novel *Centennial* was a beaver dam, for water regulation is the key to living in this landscape. Dam makers are beavers on a grand scale. Beaver dams, natural "reclamation" projects, create a landscape of rivers and ponds. The dams of engineers paradoxically celebrate both the human power to control the natural environment *and* the natural setting that is dramatized by the dams' presence. Set like great keystones, dams mark the dimensions of the canyon or valley walls. They are a testament to nature's force and power *and* to human intervention. One of the signature artworks of the 1970s was Christo's Valley Curtain, a temporary orange nylon dam suspended 1,250 feet across Rifle Gap.

Dams are an artificial geology, great stone or earthen walls creating reservoirs, artificial "lakes." Beneath their surfaces lie submerged valleys and a mysterious, underwater landscape, revealed only at its upper reaches during low water, like great bathtub rings inscribed on the former slopes. Hillsides are made into shorelines, and new land uses are born. Blue Mesa Dam, along the Gunnison, created the state's largest "lake" with ninety-six miles of shoreline and a capacity of 941,000 acre-feet. John Martin Reservoir, east of Las Animas, created eighty-six miles of shoreline along the Arkansas River. The pragmatic and the pleasurable combine as the original reasons for water projects recede in importance and the

**Photographer Unidentified**
Cheesman Dam, Jefferson County, ca. 1905. *Courtesy of the Denver Public Library Western History Department.*

**GEORGE GRANT**  Power Lines and Towers, South Side of Lake Estes, North St. Vrain Highway from the End, November 3, 1950. U. S. Department of the Interior, National Park Service photograph. *Courtesy of the Colorado Historical Society.*

ultimate effects become paramount. Thus, flood-control and water-supply reservoirs become prime recreation sites. On the shores of Dillon Reservoir is a marina that describes itself as the nation's highest yacht club. On Shadow Mountain Reservoir and Arapahoe Reservoir, high in the mountains, improbable scenes of lakes with sailboats still surprise visitors.

The state's grandest diversion project is the Big Thompson project. The Northern Colorado Water Conservancy District, the legal agency of property owners who benefit from Big Thompson projects, diverts 260,000 acre-feet of Colorado River water, supplemental irrigation for 720,000 acres in the South Platte River basin and water for urban growth and development. Water drops twenty-eight hundred feet on the Eastern Slope, creating a hydroelectric capacity of 184,000 kilowatts, originally sufficient electrical power for the entire state!

The project began in 1933, a product of Bureau of Reclamation studies and PWA funds, as a combination hydroelectric power and irrigation project. There are four main parts: Western Slope storage, transmountain diversion, foothills aqueduct, and Eastern Slope storage and distribution, including 120 ditch systems and 60 storage reservoirs. The keystone of the system is the 11.1-mile Alva B. Adams Tunnel, which cuts beneath Rocky Mountain National Park at a depth of almost one mile. West of the continental divide, Willow Lake Reservoir stores water, which is then pumped and channeled to Lake Granby, the principal reservoir and second-largest body of water in the state. (Green Mountain Reservoir on Blue River was constructed to regulate flow along the Colorado River, ensuring compliance with the state's obligations under the Colorado River Compact.) Water is lifted from Lake Granby and pumped to Grand Lake and Shadow Mountain Lake. Gravity then takes over, as water flows to the Eastern Slope—through the Alva B. Adams Tunnel, conduits, penstocks, power plants, and siphons—to storage in Carter Lake and Horsetooth Reservoir. From there water is fed to St. Vrain Creek, Boulder Creek, South Platte River, and Cache la Poudre River to be dispersed through the ditch system. The route was developed to maximize the power possibilities of the energy of water running downhill. From the Adams Tunnel to Estes Park, the drop is eight hundred feet.

In total, the system has thirteen reservoirs, twenty-seven earth and rock-fill dams and dikes, six power plants, three major pumping plants, twenty tunnels, fourteen canals, sixteen major siphons, eight penstocks, 23 million cubic yards of embankments, and 687 miles of transmission lines. It provides power for all of northeastern Colorado as well as portions of Wyoming and western Nebraska, and water for towns along the Front Range. At a much exploded scale, it recalls the sluice systems of miners. Smoothing out the troughs in the cycles of droughts, transmountain diversion projects form one of the many efforts to regularize an unpredictable climate and its seasonal and annual variation. Ironically, projects such as the Big Thompson can create new problems as the increased capacity creates new demands. Oliver Knight, in his 1956 article "Correcting Nature's Error: The Colorado–Big Thompson Project," warned, "The arid and semi-arid west may paradoxically water itself into a worse problem by superimposing industrial demands upon agricultural demands."

From the state's earliest days, the power of water has had a dual meaning, for natural forces need to be not only controlled and harnessed but also heeded. On May 19, 1864, Cherry Creek flooded Denver. The native population had warned against building along the creek, but their admonitions went unheeded. The common wisdom of not building in a floodplain was ignored, and buildings were even built in the river channel. Cherry Creek and the South Platte have flooded many times since. Dams have not proven to be a guarantee of safety. A 1921 dam break along the Arkansas River and a subsequent flash flood took 120 lives. The Castlewood Dam collapsed on August 3, 1933; on July 15, 1982, the Lawn Lake Dam broke; and the Big Thompson flooded on July 3, 1976, the state's centennial weekend, killing 139.

**ROBERT DAWSON** Dinosaur National Monument, Near the Confluence of the Green and Yampa Rivers, 1985. From the Water in the West Project. *Courtesy of the artist.*

## DOCTRINE

In hydraulic societies, the regulation of water rights is fundamental to social order, economics, political power, and the land. The law gives a look to the land. Eastern water law was adapted from English common law. Under the American doctrine of reasonable use, riparian law codified the right to use and divert water. These rights included provisions that the natural flow cannot be impeded or diminished, water quantity and quality are to be maintained, water must be returned to the stream, and water rights are retained even if the water is not used.

In California's mining districts another rule prevailed, which miners brought to Colorado. This rule became so identified with the state that it is known as the "Colorado Doctrine." Legally it is the doctrine of prior appropriation: *Qui prior est in tempore, potior est in jure* ("he who is first in time is first in right"). The law defines a priority system to exploit water resources—first come, first served—and the right is permanent. This seniority system is codified in the Colorado State Constitution of 1876, Article XVI, Section 5: "The water of every natural stream, not heretofore appropriated, within the state of Colorado, is hereby declared to be the property of the public, and the same is dedicated to the use of the people of the state, subject to appropriate as herein after provided." The constitution also asserts, "The right to divert the unappropriated waters of any natural streams to beneficial use shall never be denied." Of course, the native inhabitants lost all rights and were an exception to any prior appropriation.

Eastern riparian rights are fundamentally equable, giving equal rights to all landowners along a stream and "resting on the popular acceptance of the idea that nature should be left free to take its own course," according to Donald Worster. The rights developed in areas with sufficient water resources for agriculture and the need to regulate water power. Most historians see prior appropriation as an adaptation to the western landscape. Worster interprets it is as a reflection of changing ideas, in which nature is increasingly seen as a commodity.

To validly appropriate water rights, you must use them, and water diverted from streams must be put to beneficial use. The constitution prioritized the beneficial uses for domestic, agricultural, and manufacturing purposes, but recent rulings have extended the rights beyond economic return to include recreation and environmental protection, clearly reflecting the new water values. The state constitution also established right-of-way rights to transport water via ditches and canals across another person's land. The owner's consent was unnecessary, although just compensation was to be awarded. Water rights are property rights and can be sold, specifying amount, use, and time of year the water can be withdrawn. The maximum downstream flow can be used. There is no change in times of shortage; use need only to conform to a rule of thumb of "reasonable efficiency." The case of *Coffin v. Left Hand Ditch* (1882) upheld the state constitution and legally declared water in the West distinct from water in the East: "The common law doctrine giving the riparian owner a right to the flow of water in its natural banks upon and over his lands . . . is inapplicable to Colorado." In 1879 Colorado became the first state with official supervision of water distribution. The Division of Water Resources (in the Office of the State Engineer) now oversees seven water divisions, corresponding to the major watersheds.

Water is also regulated through a series of interstate water compacts, which govern water allocation between Colorado and states downstream. Waters flowing west are all part of the Colorado River system and are included in the Colorado River Compact, an arrangement on water rights and use between the watersheds of the Upper Basin states—Colorado, Utah, Wyoming, and New Mexico—and the Lower Basin states—California, Arizona, and Nevada. The snowmelt of the Rockies provides irrigation water for Southern California and helps fill the pools of Los Angeles, Phoenix, and Las Vegas.

The Colorado is an intensely managed river. For Marc Reisner, it is the "symbol of everything mankind has done wrong"; to the Bureau of Reclamation, it is "the perfection of an ideal." The 1922 Colorado River Compact negotiated the allocation of waters between Upper and Lower Basin states at an inflated figure of 17.5 million acre-feet per year of annual flow. Since the compact, the mean annual flow has been less than expected, salinity is higher, and the river system itself has changed under the impact of water-control projects along with farming methods and techniques of water use, welling, drilling, and pumping. In a revised Upper Colorado River Basin Contract of 1948, Colorado got 51.75 percent of the water among the Upper Basin states, but these states are not permitted to deplete the downstream flow below a certain quantity. Therefore, supplementary dams and reservoirs have been constructed to guarantee compliance with the compact and for state water needs. In recent years there has been a desire to amend the compact based on revised stream flow estimates and new Upper Basin needs such as the vast volumes of water needed for oil shale extraction.

The Colorado River Basin Project Act of 1968 authorized the building of the San Miguel, Dallas Creek, West Divide, Dolores, and Animas La Plata dam projects. The recently completed McPhee Reservoir, part of the Dolores Water Project, is the state's second-largest reservoir. Changing water consciousness has made these and other proposals the subject of continued debate. Proposals to dam the Green River, which would have resulted in the flooding of the Echo Park country of Dinosaur National Monument, were halted. The Bureau of Reclamation has proposed the Narrows Dam above Weldona. Crossing the South Platte, it would be 147 feet high and over four miles across. Two Forks Dam, proposed since the turn of the century, was recently shelved, but like many dam projects, it has had a Lazarus-like existence.

Water consciousness is changing and is reflected in a shift in emphasis in the law and in popular awareness from rights of use to the duties of use. The signs are everywhere, from the political battles over dam projects to domestic use. Homes served by the Denver Water Board have calendars posted on their refrigerators. A circle, square, or diamond denotes the days on which watering is permitted. The population growth along the Front Range communities has increased municipal water demands and competition between municipalities and agricultural uses. Public awareness and interest in stream protection for fish and wildlife, recreation, and the aesthetic dimension has developed. Any real estate property with water or even a water view is more valuable.

Unlike other resources, water is increasingly seen as a social good and a natural resource for which there is a different sense of public ownership. Water resources are subject to negotiation and compromise between diverse users and pluralistic values. Thus rivers are shared among or designated to irrigation farmers, livestock operators, Native American tribes, municipalities, industries, power companies, river runners, campers, hikers, and fishermen, all with legitimate needs to be adjudicated. The wilderness value of water is newly appreciated, and wild rivers are a key wilderness symbol. White-water recreation itself represents a significant changing symbol, since rapids are no longer proudly eliminated. They now represent a challenge to be enjoyed; one can thrill at their innate power and not their potential as power sources. In 1984, under the Wild and Scenic Rivers Act, portions of the Green, Yampa, Elk, Cache la Poudre, Big Thompson, Gunnison, Colorado, Dolores, Los Pinos, Conejos, and Piedra rivers were designated for study.

Recent community-design projects capitalize on and contribute to this water awareness in ways that recognize the multipurpose practical as well as recreational significance of water. They also emphasize the primal, poetic side of water, the pleasures of walking alongside it and observing its character—its rushing, gushing, currents, colors, wetness, states, light. Waterways in Denver, Estes Park, Pueblo, and Durango are all demonstrations of changing water values. A century ago hydraulicking, which used water to wash away valleys, was commonplace. Now Xeriscape design, the careful consideration and choice of plants that are appropriate to a semiarid landscape and that require a minimum of irrigation, is studied and practiced.

# PLANTS

## *"One laughing garden of flowers and fruit."*

IN A CENTURY AND A HALF, the Colorado landscape has been domesticated as new plants and animals have been introduced, plants and animals that now dominate the land. The patterns and products are visible everywhere: the squares of the American survey, the great circles of irrigation, the grids of orchards, the lines of furrows and fences, the engineered curves of canals. The grasslands of plains and valleys were made into fields, buffalo were replaced by cattle, nomadic cultures were displaced by farmers and ranchers.

The earliest settlers farmed the state's southern valleys. San Luis, the first permanent agricultural settlement, grew the crops of Mexico, of which it was then part: wheat, corn, beans, peas, potatoes, lentils, and chiles. Lewis Garrard's 1846 journal *Wah-To-Yah and the Taos Trail* described Hatcher's farm along the Purgatory River:

> The spot selected for cultivation was a handsome, level bottom, a mile in length and from fifty to two hundred yards in width. The gentle curving of the shallow River of Souls, its banks fringed with the graceful willow and the thorny plum, on which were affectionately twined the tendrils of the grape and hop; the grouping of the slender locust and the outspreading umbrageous cottonwoods, with the clustering currants dotting the greensward, gave a sweet, cultivated aspect to the place; while the surrounding hills, within their sheltering embrace, seemed to protect the new enterprise. The *caballada*, half hid in the luxuriant thickets, and the cows standing idly over the running waters in the quiet shade . . . served much to increase the domestic countenance of the first farm on the Purgatoire.

## FARM AND FIELD

**Photographer Unidentified** Man in Field of Oats, Routt County. *Courtesy of the Denver Public Library Western History Department.*

Gold-rush miners needed to eat, but the 1860 census showed 22,086 miners served by 175 saloonkeepers and only 195 farmers. Many miners would subsequently return to their original occupation of farming. Grubstaked by John Gregory, of Gregory diggings, David K. Wall, who had learned irrigation practices in California, established a truck garden in 1859. The *Missouri Republican* (Sept. 5, 1859) reported his garden, irrigated by the diverted waters of Clear Creek, was "the greatest curiosity about Golden City . . . some two acres in extant." The newspaper added: "It has produced as abundant a crop of vegetables as any spot seen anywhere in Missouri. Peas, beans, beets, melons, sugar-corn, radishes, cabbages, potatoes, tomatoes—in fact everything in the way of garden produce, with the exception of fruit, can be found here in comparatively astonishing quality and quantity. . . . The garden will prove the source of a small fortune to its owner this year."

The *Missouri Democrat* (Dec. 10, 1859) correspondent in Denver City offered an alternative to the gold-rush fever. "Squatting upon fine farms, and establishing ranches, is about the best and most money-making pursuit anybody can engage in out here, that understands anything of the business. Farming—raising crops of corn, wheat, oats and potatoes—will pay here next year better than even successful mining. Raising staples and fancy vegetables, such as cabbage, beets, turnips, watermelons, onions and so forth, would, with certainty, ensure men considerable fortunes by next summer and fall. . . . we have here the fairest show for agricultural and horticultural prosperity that I know of anywhere." Prudent farming, even without irrigation, could yield dramatic results. Tahosa gardens, a forty-six-acre farm near Denver, advertised a sale of fifty thousand cabbage plants, twenty thousand tomatoes, and twenty thousand beets in the spring of 1869.

Edward Bliss, in a 1861 report to the U.S. Patent Office on the Territory of Colorado, noted that there had been one hundred thousand witnesses to the fact that "The Great American Desert" was not a desert at all and that the thousands who had trekked to the goldfields knew this (most emigrants had followed the river valleys). He had seen hundreds of cattle feeding, especially in bottomlands, yet he noted, "On some of the uplands there is a tendency to bareness." Bliss was confirming the difference between the semiarid landscape of eastern Colorado and a true desert. The distinction was critical, for this land was not a "garden" as others would claim, nor a "desert," but was an area where, with proper attention,

**L.C. MCCLURE** Second Cutting, Alfalfa, Swert Ranch, ca. 1910. *Courtesy of the Denver Public Library Western History Department.*

L. C. MCCLURE   Stacking Alfalfa, ca. 1910. *Courtesy of the Denver Public Library Western History Department.*

care, and modification, an agricultural and pastoral economy could be established. Soon the Arkansas and Platte valleys began to show the signs of rural prosperity with fine, substantial farmhouses, well-fenced and well-farmed fields, and "a vast congregation of grain and hay stacks." Samuel Bowles enthusiastically reported superb melons and vegetables whose "quality, quantity and size" were "unsurpassed." He added, "The irrigated gardens of the upper parts of Denver fairly riot in growth of fat vegetables." Most successful were the irrigators. The Colorado Board of Trade could report about Greeley in 1887, "It has been demonstrated that wherever water can be got over the high lands they are wonderfully productive." Irrigation was "making a huge garden out of the once Great American Desert."

Farmers coming to Colorado brought the benefits of the beginnings of modern agricultural technology. The nineteenth and early twentieth centuries saw a successive mechanization and industrialization of the countryside. These developments augmented and increased the power, speed, intensity, and impact of very basic processes, but farming has remained, according to Verlyn Klinkenborg, "largely a matter of dragging, poking, tearing, slicing, chopping, separating, and transporting." Technology creates a look to the land, where methods leave their imprints. Much of the beauty we ascribe to the agricultural countryside is the visible evidence of these processes and stages of agricultural production, from laying out the fields to planting, cultivating, and harvesting them. The techniques create the forms.

In the early nineteenth century, the plow, the basic implement of cultivation, was modified from a wood, to a cast-iron, and then to a steel tool with interchangeable parts. With the improved plow, both human and animal labor could be expended more economically. On the prairie, sodbusting, the task of plowing primeval grasslands, necessitated an improved tool. In 1837 John Deere, an Illinois blacksmith, perfected and marketed a one-piece wrought-iron plow with a cutting edge of steel and a slick moldboard that self-scoured the sticky soils of the prairies and plains. By 1857 he was selling ten thousand plows per year.

The scythe was replaced by Cyrus McCormick's reaper, which was patented in 1834 and was followed by self-raking reapers and reaper harvesters. In the second half of the nineteenth century, the trend was to combine operations with one all-purpose tool; in the early twentieth century, autonomous self-propelled machines were created. Between 1870

**JOSEPH BEVIER STURTEVANT**  Steam Threshing Machine with Crew, ca. 1900. *Courtesy of the Carnegie Branch Library for Local History, Boulder Historical Society Collection.*

and 1900, new farm machines were perfected and came into common usage: the chilled-steel plow, the Virginia reaper, the Marsh harvester, the Pitts thresher, and most important, the grain elevator system, allowing better grain storage. Silos also made their appearance. Ensilage is the process of curing grains, grasses, and plant stalks into silage, a more nutritious fodder that results in little agricultural waste. In 1880, four men with twenty horses and an eighteen-foot cutter could harvest thirty-six acres per day. Steam-powered combines threshed and bagged grain and could maneuver over hills and rough ground. After 1900, combines were powered by stationary gasoline engines. At harvesttime, fields became the scene of nomadic factory operations. The transitions were slow, however, and horse- or ox-drawn implements were most common until the end of the century.

Manufacturing of the first successful self-governing American windmill, the Halladay Standard, began in 1854. Windmills pumped water for domestic use, for livestock, and for the railroad, which had water stations with windmills several stories tall. Wood was the most common building material, but by the 1870s, iron and steel began to be used. In 1885 the twelve-foot Halladay Standard cost $130, but the farmer also needed to erect a tower and to drill a well. By the 1890s, steel was used exclusively, with company and model names emblazoned on the wind vanes: Aeromotor, Eclipse, Kenwood (Sears), Currie, Dempster, and F&W can all still be seen across the state. Most windmills spun vertically; however, a rare horizontal wind engine was manufactured in Denver by Warren D. Parson, with "COLORADO" on the vane of Parson's Colorado Wind Engine. The windmills allowed settlement wherever a well could be dug. By the 1890s, windmills pumped water for irrigation taking the water to storage reservoirs or basins to irrigate gardens and orchards. They required no labor and pumped continuously. Before rural electrification, they generated some electricity, but most remaining windmills pump water for livestock.

Farm power progressed from human to animal to mechanical, steam to self-propelled steam-traction engines to internal combustion. The transition was not complete until after World War II. The technology of traction engines was not used extensively until the 1870s and then only on larger farms. Tractors, whose engines were most appropriately rated in "horsepower," appeared in the 1880s. The Hart-Parr Company named its 1906 vehicle a "tractor," in lieu of a "gasoline traction engine." After World War I, the tractor, combine,

and truck found their place and by the mid 1930s, horses were rarely used for power on the plains. But it was not until 1954 that tractors outnumbered horses and mules on American farms.

It is during harvesttime that the relationship of process, tool, and place is most apparent. Hay is mown, raked, stacked, and stored. Harvesting hay is an architectural, construction process: materials are gathered and processed, and great hay houses are seasonally erected. Some of these structures are even fenced, looking like grass houses standing in the midst of fields. Beaverslides, J-Hawker stackers, and derricks are still used to build vast haystacks. Before World War II only 25 percent of harvested hay was baled. Mechanical balers now deposit their product across fields in rhythmic patterns as the plunger-type baler makes rectangles while the roll-type makes cylinders that can weigh a full ton. Harold Hamil, in his high-plains memoir *Colorado Without Mountains*, wrote, "When I think of work on the ranch, the vision is mainly of hay fields . . . putting up hay was the activity that threaded through the whole summer." In just over a week, one hundred acres of alfalfa "would be transformed from a sea of waving purple to an expanse of raw, greenish-brown stubble to a dozen stacks which would reach up to form a whole new skyline."

The 1887 Hatch Act authorized agricultural experiment stations, and a year later the Colorado Experiment Station was organized at Fort Collins, to be followed by substations at Monte Vista, Rocky Ford, Table Rock, Cheyenne Wells, Fort Lewis, Austin, and Akron. Each was a testament to the distinct climatic regions of the state. At Akron, in the northeastern plains, over five thousand trees were planted to test the adaptability of plantings in nonirrigated regions.

Beyond the range of irrigation, new practices of dryland farming were employed. This farming was a new experience for settlers drawn largely from the temperate eastern states. With easy railroad access after 1870, boomers and promoters hawked the lands of eastern Colorado in an appeal to wishes, not fact. The *Elbert County Democrat* (1887) noted: "As to the possibility of farming in this region, there is no question. All that is needed is to plow, plant and attend to the crops properly; the rains are abundant." (The newspaper was owned by the Burlington townsite company.) In the treeless landscape of the plains, sod and adobe houses were common as the land was rapidly settled.

L. C. MCCLURE   Picking Peaches on Red Cross Ranch, Clifton, ca. 1910. *Courtesy of the Denver Public Library Western History Department.*

The legal mechanisms were several. The Homestead Act of 1862 offered 160 acres of land to a family head older than twenty-one. After five years of continuous residency and improvements, title was received; or title could be purchased after six months for $1.25 per acre, the more common method. The acreage and rules had been formulated based on the accumulated wisdom of eastern farming conditions and were ill adapted to the conditions of the semiarid West. For a generation, laws were passed in an attempt to come to terms with the imperatives of this new landscape. The 1873 Timber Culture Act permitted a homesteader to apply for another 160 acres if he planted 40 acres of trees, soon lowered to 10. Most applicants used this as a way to increase their holdings, although in Colorado, title to four thousand farms was granted on the basis of the act. The 1878 Timber and Stone Act also encouraged tree planting; although thousands were planted, few survived. Hundreds of orchards died in the droughts of 1892-95. According to the terms of the 1877 Desert Land Act, a settler gained title to 640 acres for $1.25 per acre if the land was under irrigation within three years. Desert lands were defined as lands that were not timbered, were not mined, and needed irrigation to produce an agricultural crop. Thus 692,794 acres were ultimately patented in Colorado (25 percent of all entries in the state).

In the 1880s, publicists, especially the railroads, proclaimed that the "rainbelt" now stretched to the Rockies, attributing this new fact to irrigation and agriculture. The theory was that changing the surface of the land changed the climate itself. Although in hindsight this was an amazing act of scientific and settlement wishful thinking, modern science also shows the interrelationships of surface acts and climate change. Between 1878 and 1886, there was more rain than usual, fulfilling the boosters' promise of an agricultural garden and the pseudoscience of rain following the plow. Inevitably, however, the cyclical patterns prevailed, and homesteaders who had expected a garden paradise found instead the disillusionment of drought, dust, fire, and grasshoppers. Most left and returned to the East.

In 1907 the International Dry Farmers Congress was held in Denver as science sought to overcome the boosters and rain salesmen. Phillip Held of Logan County reported his method of summer tillage of winter wheat. He double-disced a stubble field in spring, plowed in June to seven inches, harrowed the ground the next day to smooth ground to retain moisture, and cultivated again to break the crust and keep weeds down. Two seasons of this

L. C. MCCLURE  Planting Apple Trees Near Canon City, ca. 1900. *Courtesy of the Denver Public Library Western History Department.*

(Above) **Photographer Unidentified**  The Lincoln School, Sterling, 1942. WPA Photograph. *Courtesy of the Denver Public Library Western History Department*. Caption: "Fifth Grade boys dug vegetables, Fifth Grade girls helped sort vegetables."   (Right) **Photographer Unidentified**  Celery Pickers Working in Fields Near Denver where Pascal Celery is Grown, September 1918. *Courtesy of the Colorado Historical Society*, Colorado Highway Department Collection.

practice were needed before seeding, but the method gave four to four and a half inches of moisture in the soil. The key dry-farming idea is that land lies fallow to accumulate moisture; therefore, in the words of the agricultural historian John T. Schlebecker, the "farmer... grew one year's crop on two years' water." Dry-farming techniques all addressed themselves to maximizing moisture: deep plowing after harvest, discing after rains, leaving lands uncropped, and contour plowing. Few farmers liked "farming curved rows," according to Alvin Steinel, although the disastrous Dust Bowl experience would change their ancient habits.

By 1926 there were 11.6 million acres of dryland farming of winter and spring wheat, corn, barley, oats, rye, beans, and potatoes. Some of these lands still oscillate between cultivation and rangeland and the shifting fortunes of the boom and bust of ranchers and farmers. "Suitcase farming" made its appearance in eastern Colorado in the early 1930s. It was made possible and profitable by the cash-crop system of growing wheat, which required labor only at specific seasons, and improved road transportation and machinery. This level land was a "tractor's paradise," and the suitcase farmers, largely from Kansas, employed great cadres of combines while many local residents were still using horses. The initial boom was short-lived, for the most severe wind erosion in the Dust Bowl was at the Kansas, Colorado, and Oklahoma border. Rocky Mountain grasshoppers, known as locusts in the nineteenth century, also returned in 1937 and again in 1952. Roads were slick with their carcasses. Most eastern Colorado counties suffered over 60 percent abandonment of farms, with Baca County experiencing 80 percent. Suitcase farming returned in the late forties with huge plow-ups for wheat fields; Kiowa County acreage increased fourfold. The wheat harvest was described as "the 1948 gold rush." The inevitable bust followed in the late fifties and sixties, and 70 percent of the county lands were again abandoned.

In 1747 it had been discovered that sugar could be extracted from sugar beets, and the first factory had been built in Silesia in 1802. Root crops like sugar beets, onions, and potatoes were well suited to the soil and climate of northern Colorado. By 1870, fields yielded up to seventy tons per acre, and sugar beet factories were built in Loveland and Fort Collins. By 1926, seventeen factories were in operation, and sugar beets were the state's greatest agricultural income producer.

Many of the sugar beet workers were German-Russians living in "beet shacks." Hope Williams Sykes's novel *Second Hoeing* is set in the sister communities of Valley City, a fictionalized Fort Collins, and "Shag Town," inspired by the "Jungles," home of the German-Russian workers. Sykes's descriptions are vivid evocations of a place and time and of the labor that creates a landscape. From a distance, she presented a scene of rural comfort. "The sugar beet country of Northern Colorado; the farm-houses with gaunt cottonwoods sheltering the big red barns, high silos, outbuildings clustering close. Small fields were cut by irrigation ditches, with cottonwoods following the larger canals—that curved over the level countryside." However, on closer examination, she observed: "Separated from Shag Town by a dirt road were the stinking pulp pits adjoining the many windowed red-brick sugar factory—gray fermenting noodles of shredded sugar beets from which all the sweetness had been taken. Beyond was a high great smokestack and the black water tower on its four straddling legs."

For the workers, it was a seasonal round of stooped labor and the stink of beet pulp. Hannah, the beet worker heroine of the tale, described the fields and the work that created them: "Small beet plants, standing there three inches high, twenty inches apart, stretch, lush and glossy-leaved in long narrow rows. The dark earth was like a brown cloth marked with narrow stripe of growing green." At the end of miles of canals and channels, water was brought to plant. "She built little dams out of the wet mud at the ends of the ditches; she opened the dry ditches, and watched while the water went creeping down the dusty furrows, soaking the entire fields. it was hard, back-breaking work." The book's chapters—planting, thinning, second hoeing, and harvesting—describe the operations and purpose well. "All this work," says Hannah, "in order that just one beet, the strongest and largest, might be left to grow and flourish, one beet separated by twelve inches from its neighbor. Every thought, every effort, was put forth so that this one select plant should attain its greatest promise. All the plowing, the rolling, the preparation in the cold spring so that the ground would be as finely pulverized as machinery and human planning could make it."

The object was to grow beets with the highest percentage of sugar content, and in 1888, Colorado farmers were getting 18 percent, compared with the U.S. average of 13.62 percent.

**JACK ALLISON**
Sugar Beet Worker's Home in Colorado, 1938.
*Courtesy of the Library of Congress, Farm Security Administration Collection.*

Unfortunately, in 1919 and again in 1934, curly top virus hit the fields, leading to the abandonment of 85 percent of the acreage. Ultimately, this disaster led to the development of disease-resistant hybrid strains. Research later demonstrated that the weedy areas of dry rangelands disturbed by overgrazing were the breeding grounds for the beet leafhopper, which carries the curly top virus. Therefore, the restoration of the native plant cover and perennial grasses on the range would lower the numbers of leafhoppers and, in turn, curly top.

In the irrigated valleys of the Western Slope, commercial orchards thrived. William Pabor founded Fruitvale, the "first fruit-tract community laid out in the Grand Valley," near the current site of Fruita. The Grand Valley experienced an "apple boom" around 1895, but that too went bust and peaches now predominate. When Elbert Hubbard visited orchards around Grand Junction in western Colorado, he said, "The land is one laughing garden of flowers and fruit."

The state has always celebrated its agricultural products and culture with festivals and fairs, which were both community celebrations and promotional events. Place and product were closely identified, and the local economy was determined by the success of a crop. Grand Junction had Peach Day, Glenwood Springs had Strawberry Day, Greeley had Potato Days, Fort Lupton had Tomato Day, Buena Vista had Lettuce Day, Fort Collins had Lamb Day, and Platteville had Pickle Day. Rocky Ford's Melon Day, begun in the 1880s, was a great watermelon feast established by Senator George Swink. Swink donated eighty acres of a timber claim as the Otero County Fairground, with the stipulation that Melon Day be celebrated and free melons distributed.

## RANCH AND RANGE

Grama grasses, buffalo grass, bluestem, bunchgrass, and wheat grass—the native prairie grasses—need minimal rain. They can survive droughts and freezing, and they naturally cure into a dry feed that is edible all winter long. Prairie grasses are ideal feed for large, ruminating, grazing mammals like buffalo or cattle.

The grasses, the bison, and the plains Indian were an interdependent triad. In only fifteen years, the slaughter of thirteen to forty million buffalo (estimates vary) brought all three—plants, animals, and the Native American culture—to virtual extinction. In the 1870s, bone

**Photographer Unidentified**  Watermelon Day. 25,000 Melons Before the Feast, Rocky Ford, Arkansas Valley, ca. 1893. *Courtesy of the Colorado Historical Society.*

**Photographer Unidentified**  After the Feast, Watermelon Day, Rocky Ford, ca. 1900. *Courtesy of the Colorado Historical Society.*

pickers collected bison skeletons, which were sold for fertilizer and buttons at five dollars per wagonload, gathering bleached bones along the line of the Kansas Pacific, the first railroad to cross the Colorado plains. In one year, the Santa Fe Railroad shipped almost seven million pounds of bones and over one million pounds of hides east. By 1877, few buffalo were visible from the train.

Ranching, like agriculture, accompanied the miners. Wagon trains drawn by oxen accompanied emigrants who discovered en route that these were rich grazing lands and that "The Great American Desert" was a myth. In the first winters of settlement, it was discovered that cattle let free on the open range grew fat on the native pasture. Bliss's reports on the territory described the lands along the base of the Rockies and in mountain valleys as "great grazing and stock-raising region." For ranchers, like miners, it was also a de facto system of first in time, first in right, at least until the arrival of homesteaders. The measure of the land was an animal measure: the area an early rancher controlled was measured in terms of the distance a steer could graze from a water hole or stream, returning in a day's time.

For several decades, the now almost mythic open range reigned. It was a short-lived period, however, beginning in the 1870s and ending by 1900. The range was "open" in several senses. There were no artificial territorial boundaries, fences, or obstructions, but the land was also open in a way that appealed to the American sense of freedom, expansiveness, and not being "fenced in." The open range was a social as well as a physical ideal, which may account for its persistence. Contemporary sentiment toward the wild partakes of some of the same emotional character.

On the open range, steers grazed for three to four years and reached eight hundred to twelve hundred pounds. Cows ate thirty to forty pounds of grass per day and drank ten to twelve gallons of water. In annual spring roundups, new calves were branded as cowboys fanned out, riding wide circles in search of cattle, which were then either slaughtered or sent to the Midwest for fattening.

Cattle were also driven from Texas ranchlands to railheads on the plains. The Goodnight-Loving Trail began in West Texas, went through the Llano Estacado and New Mexico, entered Colorado over Raton Pass, and proceeded north in view of the Front Range toward

Cheyenne. Cattle needed water every five to six miles, and daily stops were about fifteen miles apart. Ideally the trail had few fords, farms, fences, timber, or rough country. Cowboys drove herds north in a great process of transhumance, the seasonal migration of people and their animals. Between 1866 and 1884, almost five million head of cattle were driven from Texas to Colorado and other range states in herds as small as one hundred and in others of thousands. In 1884 there was even a proposal for a permanent trail—five to fifty miles wide, a fenced drover's road, with streams bridged and with strategic shipping points. It was not created, for the railroad made the drives obsolete. After rails connected the state to the nation, beef were fattened at the railheads and then sent east for slaughter.

The era of the open range was doomed by several forces as "nesters" (homesteaders), irrigators, sheepherders, and railroads impinged on the territory of the stockmen. The first stock law in 1864 required stock to be herded or confined during the growing season of crops. Although largely unenforced, it demonstrated the rise of agricultural settlement. Landscape decisions again reflected conflicts between different cultural habits. Under traditional English law, it was the owner's responsibility to fence in livestock. Under Spanish law and custom, which was introduced to Mexico and on which western ranching was largely founded, the range was open to all. Thus, it was the farmer's responsibility to fence cattle out. In general, the Spanish concept prevailed. Subsequent law said it was the farmer's responsibility to protect his crops; the grazier was liable only if an animal broke through a proper fence. Currently a lawful fence is "a well constructed three strand barbed wire fence with substantial posts set at a distance of approximately 20 feet apart, and sufficient to turn ordinary horses and cattle, with all gates equally as good as the fence." Modern Colorado stockmen are not liable for livestock damage to unfenced crops or unprotected property, but they now typically fence their lands.

The Colorado Grange in 1875 resolved that all lands east of the Rockies not under ditch or within a mile of a stream should be set aside as "perpetual free pasture," but forces worked against this desire. The 1872 legislature said railroads had to pay for stock killed by trains, which encouraged the fencing of railroad rights-of-way. The public wanted fencing anyway, for cattle slowed travel. Several thousand cattle were killed annually by trains. Few were caught by the cowcatchers at the heads of locomotives.

J. F. Glidden, a De Kalb, Illinois, farmer, perfected and patented barbed wire. Glidden didn't invent it, but like Henry Ford a generation later, he mass-produced his product. Fences were a major expense for farmers; barbed wire proved to be the most economical means of fencing and, in treeless areas, the only way. Barbed wire signaled the end of the open range. Plains Indians called it "the Devil's Rope." For settlers and ranchers, it meant fewer conflicts over ownership, for one had more "positive control" over land. Barbed wire came to Colorado in 1878. John W. Prowers soon fenced one hundred thousand acres on the banks of the Arkansas as "holding pastures for steers" before they were shipped to market. Strands of different barbed wires are now collected in eighteen-inch segments and mounted in a microcatalogue, a miniaturized landscape.

Nature too changed ranching to livestock farming. There were droughts from 1885 to 1887 and blizzards in 1886. "Death stalked the plains and filled the gulches with carrion," wrote one writer. Of livestock herds, 25 to 90 percent were lost. Fencing areas became a protective measure as free methods slowly gave way to controlled grazing. The early 1880s were a cattle boom and an era of grand cattle-ranch enterprises; owners were "kings" and "barons." The Colorado portion alone of the Prairie Cattle Company (of British and Scottish ownership) was thirty-five hundred square miles with fifty-four thousand head of cattle. Hiram S. Holly's SS Ranch, near the present town of Holly, fenced in over six hundred thousand acres, including a thirty-mile stretch along the Arkansas.

The Colorado Brand Book, a compendium of cattle brands, first appeared in 1885 with over fifty thousand brands; in 1920 it listed over forty thousand brands and over thirty thousand cattle and sheep owners. Branding, like other aspects of cattle ranching, had been learned from the traditions of Mexico and Spain. From 1870 to 1880, as the state population rose from 39,886 to 194,327, the number of cattle rose from 291,000 to 809,000 head. In 1890, the year the frontier was officially declared "closed," the National Livestock Convention was held in Denver, accompanied by a great barbecue of buffalo, elk, bear, antelope, beaver, sheep, and possum. It was an elegiac feast.

There were conflicts over grazing and water rights, conflicts between farmers and stockmen, and lands were overgrazed. In 1869, merino sheep were introduced to Colorado. By 1885, the total wool "clip" was eight million pounds, about seven pounds per animal.

L. HOLLARD   Trail Herding, ca. 1890. *Courtesy of the Denver Public Library Western History Department.*

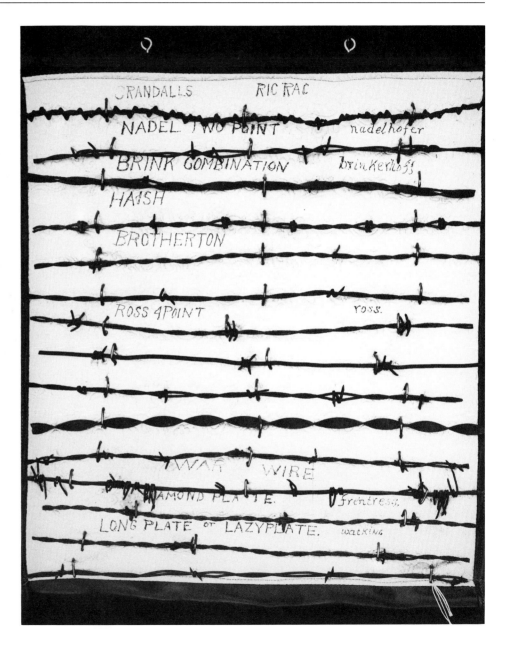

**ROBERT DAWSON**  Barbed Wire Display in the Animas Museum, Durango, 1986. *Courtesy of the Animas Museum, La Plata County Historical Society, Durango.*

Sheep graze closely and tramp down the earth, leaving the ground almost bare. They graze in rockier areas and foothills areas unattractive to cattle; therefore there was some natural division of grazing lands, but conflicts were inevitable. After the establishment of a system of Forest Reserves, later to be the National Forest Service, there were also conflicts with the government as a regulator of grazing rights. The Taylor Grazing Act of 1934, named for Colorado Springs Congressman Edward T. Taylor, the chair of the House Appropriations Committee, established grazing districts in eleven western states and a system of grazing permits. It led to the creation of the Bureau of Land Management in 1946 to manage rangelands.

Raising cattle has changed from the open range to ranching to livestock farming. As early as the 1870s in Weld County, stockmen located near feed-producing areas and irrigated lands. Cattle and sheep were fattened on cheap and nutritious beet tops and beet molasses. East of Greeley lies the Montfort feedlot, now operated as a division of Con Agra, "the only company which operates across the entire food chain." The feedlot has one hundred thousand yearling steers in one-acre pens, with four hundred cattle per pen. The feedlot is an incredibly concentrated grazing land of powerful ratios—four hundred cattle per acre versus fifty acres per animal on arid rangeland. It is an intensified landscape. A mill at the feedlot processes the grain for the cattle, and all feeding is computerized. Cattle are fed 100 to 130 days before reaching an ideal weight of 1,150 pounds for steers, to yield "quality carcasses of 725 lbs." About half of the animals are used for meat, the rest for by-products; even the runoff water is collected in holding ponds, to be recycled and used to irrigate farmland.

Ranches appear pastoral, but they are the scenes of hard work. One author calls the beef cow a half ton of "stubborn, complicated and highly perishable merchandise." In 1890 Colorado had 16,389 farms and, by 1910, 46,170, reaching a high of 59,956 in 1930. There are now only 26,600 farms in the state, averaging 1,237 acres each. A century ago, most Coloradans personally knew farming and farm life, and a half century ago, most people at least had a grandparent or a family member still on the farm or ranch. Now few people have direct experience working the agricultural landscape. The rural landscape is near but distant: for most, it has become part of the scenery and history.

**B. H. GURNSEY** The Largest Cottonwood Tree in Colorado, Fifth Street, Pueblo, 1879 (detail). *Courtesy of the Denver Public Library Western History Department.* Caption: "Gurnsey's Rocky Mountain Views, Scenes On the Line of the Denver and Rio Grande Railway." Known as "The Old Monarch," this tree was 380 years old when it was cut down in 1883. "In 1850 there were 36 persons massacred near this tree by indians. Fourteen men have been hung from one of its limbs. The first white woman who died in Colorado was buried under its branches." Text from back of similar print.

## HOUSE AND GARDEN

Early Colorado explorers and visitors, all citizens of a nation of farmers, commented on the plants in the landscape and on the agricultural potential of the land. Plants and trees especially were indicators and signs. George Ruxton, along the Arkansas River east of Bent's Fort, observed: "Nothing meets the eye but a vast undulating expanse of arid waste; for the buffalo grass, although excellent in quality, never grows higher than two or three inches, and is seldom green in color; and being thinly planted, the prairie never looks green and turf-like. Not a tree or shrub is to be seen, except on the creeks, where a narrow stamp of unpicturesque cotton-wood only occasionally relieves the eye with its verdant foliage."

J. C. Frémont said the cottonwood "deserves to be called the tree of the desert—growing in sandy soils, where no other tree will grow; pointing out the existence of water, and furnishing to the traveller fuel, and food for his animals." Cottonwoods grow in moist ground by streams, water holes, and old buffalo wallows. The Oregon and Santa Fe trails "went from one grove to the next for water, fuel and shade. . . . these cottonwood groves were the wayside inn, the club, the church, the newspaper and the fortress when the wagons drew up in a circle beneath the bows [sic]." One such grove along the Arkansas, Big Timbers, had been an Indian encampment and became the site of Bent's Fort. Later, places such as McBroom's Grove along Bear Creek near Morrison became renowned sites for picnics and July 4th celebrations.

Cottonwoods (*Populus deltoidea*) are the poplars of creeks, whereas the uplands of Colorado are distinguished by aspens (*Populus tremuloides*), the trembling poplars of the mountains. Aspens, or quaking aspens, are environmental and cultural indicators, a successional forest species invading sunny, cleared sites. Tolerant of poor conditions, aspens improve soil quality and are a sign of long-term regeneration. Their leaves do tremble and quake, making a delicate musical sound. Aspens can be found almost anywhere in the mountains but grow mostly at an elevation between 8,000 and 10,000 feet. They are golden-yellow in the fall and shimmering in the summer, yet it is an irony that these sites of great beauty are evidence of disturbance and sometimes indicate areas despoiled by mining, logging, fire, and grazing.

Planting is a domesticating act, a way of fighting the wild. In 1864-65, citizens of Boulder and Denver transplanted cottonwoods from the river bottoms to town. Since then, millions

**KENNETH HELPHAND**  Arkansas River Valley Ranch, North of Salida, 1989. *Courtesy of the author.*

of trees have made oases of Colorado towns, ranches, and farms. Main Street, the commercial heart of small communities, is treeless. The workaday world typically is not greened. However, trees along streets and on house lots ring most town centers. Domestic planting, the world of house and garden, forms the predominant green structure of communities. The seemingly modest choice of plant species for front yard and backyard reveals a fundamental interaction between culture and nature. Our plant choices reflect a cultural heritage, perhaps a species heritage, in the preference for green itself and for the expanse of lawn. Preferences are also born of familiarity, the comfort of the known.

The most common Colorado garden is, not surprisingly, the "American" front, presentational lawn-yard and the workaday space of the backyard. The lawn of water-loving Kentucky bluegrass is not native to Colorado. In 1988 the *Denver Post* columnist Ed Quillen, tongue in cheek, proposed outlawing private lawns. It would save water, money, and time, and he saw lawns as a wasteful habit. But the front lawn is the American garden, the base on which we build our home, the true frame of our dwellings. Representing domesticity, it is an atavistic attachment to the land and a reminder of a rural past. Gardens are made from nature, but they are culture.

There is a Colorado landscape aesthetic of form, pattern, texture, and color. It includes mountain silhouettes and distant vistas, piñon pines on red slopes, golden aspens in wavy bands, wildflowers in meadows and rock crevices, the soft band of trees along a stream, and rocks. In popular and personal form, these elements are miniaturized and condensed in residential landscapes. People plant the garden ABC of aspens, boulders, and the Colorado blue spruce, all regional symbols united by the ground of the American garden lawn.

The unit of house and garden is the basic building block of the settled landscape. The typical rural home is set in a planted grove, the conscious creation of a sense of enclosure, a protected microclimate shielding winter winds and snows and providing cooling shade. Seen from distant fields, the tree-surrounded home is a marker at day's end. On the plains, trees are not a symbol of nature but of civilization. They are the sign of human presence. Thus travelers, after a long trek or drive, mark a settlement by trees on the horizon. Towns such as McClure appear as green, tree-walled plats, with only a water tower punctuating the frame. You pass through the green screen separating field from town. These are not "street"

trees but "house" trees, trees on house lots, trees planted to provide shade for dwelling, work, school, and park and a respite from the harsh sun and the wind of open fields. The trees are a windbreak, a marker, a shelter, and a sign.

Gardens are created by acts of the mind as well as the hand. In Colorado, the discovered garden is the most important, where one finds order, pattern, magic, and comfort in the natural world. W. H. Kidd, en route to the Gregory diggings in 1859, reached the summit of a road about seventy miles north of Pike's Peak. He was astonished by the vista: "Instead of barren and desolate mountains, we were greeted with beautiful garden plots, abounding with gooseberry and current bushes, and orchards of wild plum and cherry. Parks of various dimensions and shapes, dotted with pine trees and fringed with grassy openings, all planned and planted in nature's own beautiful order." Thousands still seek out nature's "gardens," as the path of spring is traced in annual treks up mountains trails in quest of alpine wildflowers. There is an admiration of the adaptations of the plants, emerging out of the snow, growing in protected cracks, and bursting in color where least expected.

Plants may be characterized as native or naturalized. Native plants are indigenous to an area; those that are naturalized take root, grow, and thrive without human intervention. Naturalized plants appear as if they belong. Designers would say they "fit," like clothing. An awareness and an appreciation of what is native and naturalized to Colorado's life zones need to be cultivated and nurtured.

What kind of gardens do people willfully make in this landscape? The garden is the idealized landscape, the landscape miniaturized and tended, a place to rest, collect, use. It is the stage for domestic life. In 1868 a visitor described Colonel Craig's Hermosilla Ranch, along the Huerfano River, with its mansion, herds of cattle, sheep, and horses, and hundreds of cultivated acres. "A splendid grove of tree is fast enveloping the house; acequias [irrigation ditches] meander through the grounds in every direction; a reservoir supplies a fountain in front of the broad piazza and clematis and virginia creepers are trained to the cornices. A couple of thousand fruit trees, vines and shrubs have been received, and will add to the attractions of a modern Eden; a veritable oasis in the Great American Desert."

A new garden ethic is struggling to root itself within Colorado, an ethic based on choosing local plants (or, more often, the plants "volunteer" and choose the place), native and

**L. D. REGNIER**  Home, May 1902. *Courtesy of Rodd L. Wheaton Collection.*

naturalized. Bright sun, low precipitation, low humidity, dry winds, and extreme differences between temperatures of day and night all deprive plants of life-giving water and present a horticultural challenge. Rocky Mountain vegetation is typically pale-green, rigid, small-leafed, gray to blueish. It has hairs, wax surfaces, or scales and extensive root systems—all ways to conserve water. The Xeriscape movement promotes water conservation and an appreciation of indigenous landscapes and is slowly changing the persistent ideals still adapting to the Colorado environment. Demonstration Xeriscape gardens are found at the entry to the Fort Collins City Hall and at the Denver Water Department. It is possible to have a green city and to conserve water.

Gardens also miniaturize, map, and model places. Gardens are collections, not only of plants. The alpine rock garden is one of the most common forms of plant collections. Early homesites, especially in Denver and Boulder, often incorporated rocks into gardens. At the turn of the century, Darwin M. Andrews's Rockmont Nursery in Boulder was the first to offer native shrubs, and the landscape architect S. R. DeBoer built the first public rock garden in Denver as part of the Sunken Gardens on 8th Street and Speer Boulevard.

Touring the side streets of the state reveals wonderful acts of imagination and interpretation. In Olathe, topiary spells out "HI THERE" to passersby. The garden of Sterling's historical museum includes petrified rock and historical artifacts, along with an exhibit of native prairie grasses. There are also the ubiquitous piles of collected mountain rocks, ore buckets, outmoded farm implements, and wagon wheels. These artifacts, placed for all to see, are modest monuments and shrines in nostalgic honor of the previous generations' struggles with the land. Skulls, antlers, and statues of wildlife are displayed. In Pueblo, a front yard is a gravel-and-rock model of the Rocky Mountains and plains, with a blue Arkansas River flowing across the surface. On closer examination the mountains, made of painted gravel, show veins of gold!

# PART THREE

*Connections*

*Settlement*

*Visions*

**HARRY H. BUCKWALTER**  Engine #3 of the Cog Railroad up Pikes Peak, ca. 1900. *Courtesy of the Colorado Historical Society.*

# CONNECTIONS

*"At every turn something new and unexpected comes into view."*

THE LINES OF HUMAN IMPRINT are pervasive and powerful. They mark routes of movement, corridors of activity, and channels of communication. This is the landscape of transportation, the movement of people, goods, and ideas: trails, tracks, roads, airways, pipelines, and satellite dishes. Transportation environments are not only physical artifacts and connections. They are also community spaces, short-lived unions—often of strangers—connected by a common passage. There is the life on the trail, the camaraderie of the railroad coach, the shared road of commuters and vacationers, the conversation on the ski lift.

Animals leave tracks, ephemeral traces, the mark of a recent passage. A tracker knows how to read the signs for indications of the animal's size, speed, condition, and direction. So too one can "track" the patterns of human movement in the landscape, patterns equally telling of our condition. Modern technologies have accelerated the speed and power of movement, and new equations of time and distance are reflected in landscape patterns. Our traces are visible in corridors of movement whose tracks persist long after activity has ceased.

Access and accessibility, bringing people and goods to places, has been a key theme in the Colorado's history. The state has woven together its regions while simultaneously linking itself to national networks. The result has been the diminution of isolation and a decline in the differentiations between urban, rural, and wild lands. In 1846, along the Oregon Trail, Francis Parkman noted, "Ten feet from the wagon traces the wilderness began and civilization ended." Now satellite dishes receive signals from everywhere, and one can carry a cellular phone into the wild and be connected to virtually any site on the globe.

## ACCESS

Initially, there was the exploratory process, followed by the selection and creation of routes. Lands were "opened up," made ready for exploitation, extraction, development, and settlement. Colorado is replete with bold access initiatives. At the cliff dwellings of Mesa Verde, now reachable by ladders and steps, the original handholds and steps, carved into precipitous rocks, are still visible. Passes were ascended by sheer willpower, railroad tunnels were bored through mountains, water was brought across the divide, roads reach to the top of Mount Evans and Pike's Peak, and ski lifts ride into the clouds.

Each transportation mode has its own character, imperatives, and engineering requirements, which shape the land and our perception of the territory traversed. Modern paths across the land have progressed from the rigid fixed tracks of the railroad to the more fluid and flexible routes of the automobile, to the open possibilities of flying. The functional differentiation of transportation paths implies particular types of travel experience. The vehicle, the route, the sequence, and one's companions all structure the character and quality of the experience.

**WILLIAM HENRY JACKSON**
Canon of the Rio Las Animas, ca. 1881. *Courtesy of the Colorado Historical Society.* Jackson, who was trained as a painter and retoucher, took great liberties in "improving" his later photographs. Note that the left and right sides of the canyon have been heavily painted.

## TRACKS

Before the Civil War, the federal government surveyed and planned routes for a transcontinental railroad, identifying ten possible routes across the Rockies. The mountains proved to be a formidable barrier, however, and the first passage, completed in 1867, bypassed Colorado and went through southern Wyoming. Colorado soon connected to the national network via feeder lines, but the history of its railroads is one of confrontation with the mountains. In 1870 the Denver Pacific linked the 106 miles from Denver to Cheyenne, and the Kansas Pacific also extended to Denver. In just over a decade, lines ran from Trinidad and Pueblo north to Longmont and Greeley, west into mining districts at Black Hawk, Central City, and South Park, up the Arkansas through Royal Gorge and toward Cañon City, up the Rio Grande as far as Wagon Wheel Gap, and to Silverton, Leadville, Rock Creek, Dillon, Gunnison, and Crested Butte. Narrow-gage railroads made this rapid development possible. Eventually the lines probed to, and often dead-ended at, Creede, Somerset, Aspen, Westcliffe, Graymont, Eldora, Ouray, and Ironton, wherever minerals were to be found.

The railroad is constrained by terrain, of necessity taking the path of least resistance. The "line" cuts its own route through the landscape, deviating only when necessary and

**WILLIAM HENRY JACKSON**
Boulder Canon Near Castle Rock, 1873.
*Courtesy of the United States Geological Survey, Denver.*

modifying the land to fit its rule, regulations, and engineering limits. The ideal is a virtually level roadbed. Tracks initially went across flat lands and up river valleys, but how to get into and over the mountain barrier? Narrow-gage tracks, 3 feet across versus the standard gage of 4 feet 8 1/2 inches, were a response to steep grades and confined canyons. The upper range for the grade of standard-gage trains is 4.5 percent, but it is 7 percent for narrow. The narrow-gage Denver and Rio Grande (the D&RG) became the most extensive rail line in the state.

It is easy to miss the efforts, the public works, that allow the railroad a smooth and easy flow. During construction, grading gangs evened out the land to its slightest wrinkles and cracks. Most often the rail line sits on its own embankment, a terrace over the land, making thousands of crossings of streambeds, gulches, washes, depressions, and roads. The system necessitates culverts, cuts, viaducts, bridges, and tunnels. Iron rails, wooden ties, and earth for the bed on which the rail "rests" were brought together. To keep the line going required more wood and a continuing supply of coal and water for fuel and steam. The spacing of water tanks measured the land and marked the railroad's path. Despite the abandonment of hundreds of miles of tracks (mostly narrow-gage), the remains of railroads are everywhere, in the ruins of roadbeds, bridges, trestles, abutments, piers, and ties.

The straight lines from the east curve and rise to cross through the mountains. At certain points they loop over themselves in spiral fashion. There are switchbacks, even the geared notches of the Manitou and Pike's Peak Cog Railway, which rises 7,500 feet in 8.9 miles to the summit at a maximum grade of 25 percent. Emily Faithful, an Englishwoman visiting in the 1880s, wrote in her *Three Visits to America* that the man who planned the Denver and Rio Grand Railway seemed "to have lassoed the mountains and raught them in a tangle of coils." Tracks over Kenosha Pass to South Park at 9,950 feet, in 1879 the highest rails in the continent, had 560 curves.

Gorges and canyons were spanned by bridges, dramatizing the state's geological wonders, as canyon sidewalls were tunneled and railbeds were blasted along riverbeds and along cliffs through Royal Gorge, Gore, Glenwood, and Ruby canyons. Almost immediately Colorado's engineering achievements were converted into attractions. The Georgetown Loop Railroad, from Georgetown to Silver Plume, first carried ore when it opened in 1882, then passengers, and soon tourists. Demolished in 1937, the Devil's Gate High Bridge of the Georgetown

Loop was reconstructed and the line reopened as a tourist attraction and historic landmark in 1983! At nine passes, railroads actually crossed over the continental divide, but for year-round routes, tunnels were a necessity. The Alpine Tunnel beneath Altman Pass, at an elevation of almost 12,000 feet, connected Gunnison and Buena Vista. Begun in 1880, it was often closed due to heavy snows and was abandoned in 1910 after a cave in. The 6.2-mile Moffat Tunnel beneath James Peak was the longest railroad tunnel in the Western Hemisphere when completed in 1927. It made possible direct transcontinental rail travel through Denver for the first time and helped open up western Colorado.

Electric and street railways, interurbans, also knitted communities together. Beginning in the 1890s, interurbans operated in the Cripple Creek district, Grand River valley, Trinidad, Fort Collins, Greeley, Gunnison, Montrose, Durango, and Denver. Interurban routes began to be abandoned in 1915, under pressure from automobiles and buses, since there was little support for urban mass-transit systems. But for a time, the Denver Tramway Company (DTC) controlled the Denver and Intermountain (D&IM) and the Denver and Northwestern (D&NW), which ran until 1958 along with DTC streetcars. The Denver and Interurban (D&I) operated from 1908 to 1926 between Denver and Boulder, Fort Collins, and Greeley.

Rails are part of transportation corridors. In the Arkansas River valley there is a parallel triad of river, road, and railroad. The railroad through the river's Royal Gorge was originally practical, but it soon became an attraction, with tours in open flatcars. Emily Faithful wrote:

> Mr. Ruskin's heart would indeed have ached to see the solemnity and majesty of this weird ravine desecrated by the noisy, ugly, puffing locomotive which drew our trains through its mystic shade by the side of the river, under the giant cliffs 3,000 feet high, that seemed to frown on its intrusive presence, and even to threaten its puny form with destruction! The giant of the 19th century—the ogre who, while he brings these lovely places within ordinary reach, spoils their picturesqueness and destroys their solitude—is gradually asserting his sway throughout the wild districts.... the spirit which animates Wall St. asserts itself in the wild canyons of the Rocky Mts.

**GEORGE BEAM** Wolcott, Milepost 318 Looking West Across 60,000 Yard Fill, August 8, 1927. Laying New Track for the Denver and Rio Grande Railroad. *Courtesy of the Colorado Historical Society, Denver and Rio Grande Collection.*

In 1929 the Royal Gorge Bridge was built overhead, spanning the gorge. Why was it built, when the bridge does not lie along a significant transportation route? It is an exceptional engineering feat, allowing one to cross over the gorge and then return. A century after Emily Faithful's visit, one can cross the bridge in a car or on foot (it is often as crowded as a downtown street), take a funicular down 1,550 feet to the river, raft the gorge, or cross it in a cable car. There is a kiddie railroad as well. The gorge, proclaimed to be GORGEOUS on billboards, is now a theme park, where the means of movement become paramount. Access, which began as pragmatic, has become recreation, and difficulty, now controlled, has become entertainment.

In 1926, 116 passenger trains went through Denver each day. Only the daily whistle blasts of steam engines on tourist rail lines remain as extremely modest reminders of the past whistles, air pumps, brakes, smoke, and steam. But Harold Hamil, growing up along the Union Pacific line, remembered: "The passing trains were a part of each day and night, we saw them, we heard them, and when we were anywhere near the tracks we smelled the pungent smoke that gushed in opaque columns from the locomotive stacks."

**CHARLES GOODMAN** Great Snow Slide and Snow Tunnel on Wagon Road Between Ouray and Silverton, July 6, 1888. *Courtesy of the Colorado Historical Society, State Highway Department Collection.*

## ROADS

Across the plains routes of roads, rails, poles, and wires run parallel. Each line follows the other, reinforcing the original corridor, so that the initial choice of route persists. We go where others have blazed the trail and marked the route. Driving along, we easily neglect the road as the most obvious and dramatic landscape determinant. With the gold rush, connections between mines and markets were rapidly established. A letter in the *Missouri Democrat* (Dec. 1, 1859) said, "Laying out and making fine wagon roads, from various points and towns in the valley, up through the various passes and sections of the mountains, seems to be the order of the day." Territorial charters granted the right to build roads and collect tolls, but most of these early proposals were constructed only on paper. However, there were toll routes from Denver to Central City by 1860 and toll routes and stages over Kenosha Pass by 1865. The practice continued after the rushes. Otto Mears built over three hundred miles of road in the San Juan Mountains, including his Million Dollar Highway between Ouray and Silverton. However, most high-country roads came "into being simply by driving a wagon along and lining the wheel tracks with stone markers," according to Marshall Sprague.

**REX MOLLETTE**  Streetcar, Main Avenue, 1911 (detail). *Courtesy of the Animas Museum, La Plata County Historical Society, Durango.* Photograph shows repair crews working on damaged trolley track after the flood of October 1911. Debris from the flood can be seen on the street.

Routes were cut through timbered areas, roads were graded, streams were culverted, log-pole bridges were erected, and logs were laid as corduroy roads across swamps and soggy areas. This has happened for thousands of miles and is indicative of the yard-by-yard effort of digging, cutting, plowing, and marking that domesticates a landscape.

May 1899 saw the first car in the state, an electric automobile. The first car for sale was a $750 Locomobile, a steam-powered vehicle. Of course, it was the gasoline-powered internal combustion engine that would win out. In 1901 W. B. Felker and C. A. Yont drove a Locomobile to the top of Pike's Peak. A year later Denver had two hundred cars. In this early period the automobilist was an adventurer, challenging rutted, muddy, or icy roads in open-topped vehicles. In response to these conditions, the Colorado Good Roads Association was formed and, along with the Colorado Automobile Club, helped generate the momentum for a State Highway Commission, created in 1909. Communities celebrated roads linking themselves with other places. Greeley and Golden erected entry arches along their main street. Golden's is the rare survivor.

Henry Ford's 1908 Motel T signaled the beginning of the twentieth century as the age of the automobile. The car became fundamental to our culture and landscape, a vehicle for the masses, expanding the size of communities, dispersing the pattern of settlement, enabling suburbanization, and transforming our way of life. The landscape in which most Coloradans now dwell is an automotive landscape. Roads, with their easements, are typically a third of the land area of communities, and highways are our largest construction projects. The ubiquitous often goes unrecognized, but road size, surface, and signs are governed by national standards. Any stretch of street is curbed and guttered, sidewalks parallel the roadbed, and automobile access is provided for each lot. We rarely think of it, but when we are outside, cars are almost always in our view, either parked along the street or moving by. On the roadside we expect to find the necessary directional information in the form of signs on every corner telling us where we are and leading us to our destination. We expect services for the car and the traveler to be readily available. We expect to be able to transact business and buy products and entertainment from behind the wheel or in places designed for easy automobile access. At the domestic level, cars enter driveways and are stored in garages attached to houses. For most new homes, the entry "walk" is the "drive"-way.

In 1909 the idea of the Lincoln Highway, a transcontinental highway trail, was proposed at a state motorists' convention in Pueblo. It was suggested that this be a tree-lined route with representative trees of the states it passed through. The Lincoln Highway, later U.S. 30 and then I-80, just skirts the state's northeast corner; however, many trails would soon cross the state, for what was impossible for the railroad was realizable for the automobile. By 1916 the federal government began to subsidize state road construction, and for the next sixty years highway building would be an omnipresent activity. The National Highway Association and the Good Roads organizations promulgated a system of national roads, which evolved first into a system of national "trails" and then national and state numbered routes, culminating in 1955 with the Interstate Highway System.

During the 1920s, several hundred marked automotive trails crossed the nation. Through Colorado passed the Rocky Mountain Highway, National Park-to-Park Highway, Victory Highway, Roosevelt Midland Trail, Albert Pike Highway, and the National Old Trails Road following the path of the Santa Fe Trail. The best-known trail was the Pike's Peak Ocean-to-Ocean Highway, which characterized itself as "The Appian Way of America." The P.P.O.O. route went through Burlington, Limon, and Colorado Springs. There it branched to Buena Vista, Leadville, and Glenwood Springs or toward Cañon City, Salida, and Montrose, with the branches merging at Grand Junction. The P.P.O.O. tourist guide reminded its users, "At every turn something new and unexpected comes into view." Trails were marked with painted symbols on telephone poles. The symbol for the P.P.O.O. was a red-and-white band.

Driving through the mountains was a difficult proposition. Courtney Cooper offered urban dwellers advice on driving in the high country: "In the city you look for the signal of the traffic patrolman, watch the man ahead of you, and keep an eye open for what may come upon you from the intersecting streets. That makes three things for the eye to watch. In the mountains there is only one—and that's the road. But, it must be stated, that the road is a jealous thing, and one which requires constant attention. You don't just give it a glance now and then; you watch it!" In the 1930s, most improved roads outside of main city streets were still graveled, and blacktop was a novelty. Notable exceptions included paved roads to the Denver Mountain Parks, up Pike's Peak, up Lookout Mountain, and to Rocky Mountain National Park.

**CHARLES S. LILLYBRIDGE**
Bicycling Along the South Platte River, Denver. *Courtesy of the Colorado Historical Society.*

(Left) **Photographer Unidentified**   Montrose, Preparing to Pave the Streets, 1919. *Courtesy of the Denver Public Library Western History Department.*   (Above) **HAROLD CORSINI   Road Repair Operations, U. S. Highway 50, June 1949.** *Courtesy of the Photographic Archives, University of Louisville, Standard Oil of New Jersey Collection.*

I-76 follows the South Platte, but I-25 and I-70 make a great cross over the state and are the major routes in the cardinal directions. The interstate system was conceived as a limited-access, almost supranational network linking all parts of the country. This has certainly been accomplished, but the system also transformed and continues to create landscapes, as this national scheme has profound local implications. I-25, I-70, and their appendages of bypasses and arterials, particularly in metropolitan Denver, accelerated urbanization and sprawl from Colorado Springs to Fort Collins. Roads at this scale have their own imperatives, to which other developments defer. The difference is apparent at sites such as Loveland Pass, where since the completion of the 8,960-foot Eisenhower Tunnel in 1975 (aptly named for Eisenhower's perhaps most lasting achievement as president, the Interstate Highway System), a pass previously traversed at a careful twenty miles per hour is now cut through and barely noticed at fifty-five miles per hour. The highway continues over Vail Pass, named for Charles D. Vail, the chief engineer of the State Highway Department, who had originally redesigned the Loveland Pass road for cars. The imperative is most dramatic in the continuing struggle to channel the highway's final segment through the narrow walls of Glenwood Canyon with a divided highway almost hanging on the canyon wall. The segment is scheduled for completion in 1993. The I-70 embankments are the state's largest tailings. As I-70 cuts through the foothills, there is a butcher-shop revelation of the bones and sinews of the landscape as the stratigraphy is revealed and the banded layers are marked off and labeled.

The speed of the road seems to reduce the distance through the land. I-70 has made the mountains more accessible as lands, especially in the one hundred miles west of Denver, have been placed within daily and certainly weekend range, effectively increasing the city's recreational "watershed." The most dramatic landscape impacts are the development of the ski and resort communities that now lie astride the roadway. I-70 has become a winter ski corridor facilitating the growth of Vail, Keystone, Beaver Creek, and Breckenridge. These sites are now also linked to an air system. Skiers arrive at Denver's Stapleton Airport from throughout the nation, then hit the road or board another plane to their destination. Highways are already being planned to lasso Denver's new airport into the road network, and a new city of support services is destined to develop on the airport's flanks. Smaller airports in Aspen, Telluride, and Durango have a similar impact at the local level, bringing remote locales into home range.

America has been called a "car culture," and the state has many varieties. On Friday night in Trinidad, cruisers reassert the primacy of Main Street. In rural communities such as Antonito, Saturday afternoon cruising is essential as a way of gathering. In the metropolitan areas, the commuting range has expanded. Park-and-ride locations are found deep in what were recently rural communities, as commuters overlap in habitation with second homes, country cabins, and ranchers and as pets compete with wild animals. Alleys are still found in most Colorado towns. The alley is a rural remnant, a place typically unpaved, a reminder of the thin veneer of pavement put atop the landscape. Filled with weeds and puddles, it is a more wild place—ask any child who seeks the alley as the preferred route home.

A look at the state highway map shows mountains and mesas encircled by paved roads, with dirt roads entering the mountains and remote grasslands like probes into an alien country. Access trails, into these areas bring the backcountry to the fore. Four-wheel-drive vehicles and jeep tours now make what was available to few an adventure for many. Previously, winter limited landscape access, but technology overcame snow and ice: powerful locomotives now clear railroad lines, roads are paved and plowed, and four-wheel-drive vehicles, snowmobiles, lifts, and gondolas negotiate winter conditions.

There are also new vehicles, technologies, and forms of "exploration." They do not always mark the land in the manner of rails or roads. Some leave tracks, but others leave no traces at all: jeeps, ATVs, dirt and mountain bikes (a sport pioneered in Colorado), river rafts, skis, and hang gliders. Taking people to new landscapes, these technologies emphasize personal experience and skill in knowing the vehicle, terrain, qualities, and characteristics of the route and changes along the way. The self is explored through the place, as accessibility is increasingly individualized, even tailored. There are more choices for how to get to more places, and land managers are confronted with the dilemmas of a democratic landscape, of restricting numbers and types of access. Perhaps some places need a kind of historic access status, allowing entry only by the means first used. Accessibility is not only a spatial concept but also a temporal one. Trails and roads can take travelers back in time as well as space. Seeing the ruts of pioneer routes or following the explorer's path provides a visitor not only with a connection across space but also a linkage through time.

**Photographer Unidentified**   Photographs Showing Road Conditions in San Luis Valley, March 1931. Page From State Highway Department Scrapbook. *Courtesy of the Colorado Historical Society, Colorado Highway Department Collection.*

Roads are more than just paths for vehicles. Hamil's plains memoir, describing the road outside Proctor around 1910, reminds us of a child's intimacy with landscape, where naming the world is important and there is a closeness with the surface of the land, where inanimate elements have character and can be, as he notes, friends. The landscape is fundamental to human development, and we each have our own environmental autobiography.

> This was a quite ordinary stretch of road by most people's standards, but it was a miracle mile, almost, to youngsters for whom it was the principle outlet from home, the route to Proctor, the first leg of any trip to the outside world. We trudged the length of these two lines of barbed wire at least a thousand times on the way to and from school. It was the route of our first horseback ride, our first trip at the wheel of a Model T. . . . But it was on foot that we developed familiarity with every detail. We favored the soft, alkali-flecked dirt shoulders in summer when we barefooted, and we sought out the widest wagon ruts when we walked in clumsy overshoes through the snow and ice of winter. We made friends of every bridge and culvert, every gate, every misshapen fence post, every break in the wire, every clump of weeds. There was the "big ditch," which wasn't big at all, but considerably bigger than the "little ditch." There was "pigpen ditch" that we crossed on the outbound trip. . . . The slough and the alkali spot were major landmarks, and about midway between them was a fence post that had been designated as the half-way point between home and school.

## SIGHTS AND SITES

In its formative years as a territory and a state, Colorado experienced an unusual coincidence of visitors. Tourists arrived virtually simultaneously with the state's explorers and settlers. Tourism democratized the hardships of the explorer and coupled it with the zeal of the pilgrim. Tourists are on modern-day quests, following paths laid down by others and set on recapturing something lost—often embodied in landscapes. They return home with evidence of their journeys, from snapshots to souvenirs. In Colorado, the "Old West" was visited and mythologized in the process of its transformation. In 1870, cattle roundups brought sightseers and excursion trains.

**JOSEPH BEVIER STURTEVANT**   A Texado Snow Ball on Line of C & N Railway, July 19, 1899.
*Courtesy of the Carnegie Branch Library for Local History, Boulder Historical Society Collection.* Texas school teachers, who vacationed at Boulder's Chautauqua, on a "Texado" excursion to mountain snowfields.

Mass tourism is one of the quintessential modern landscape experiences. In the touristic landscape, sights—attractions—are identified and created, access is facilitated, accommodation is provided, and activity is directed. A tourism "industry" has developed that accomplishes all this with increased sophistication in the recognition of society's diverse desires and needs. In addition to their baggage, tourists carry a set of expectations crystallized as images of places and experiences. These images are derived from both reality and myth and are promulgated through promotions of all types. Places are then designed and orchestrated to fulfill those expectations. The industry often revives old ways as "new" attractions. After a period of time, the old is new once again and ripe for rediscovery.

Horace Greeley, on visiting Gregory Gulch in 1859, was aware of this compression of time and activity, the confluence of pioneer and tourist. He accurately foresaw the future:

> Mining quickens almost every department of useful industry. Two coal pits are burning close at hand. A blacksmith has set up his forge here, and is making a good thing of sharpening picks at 50 cents each, a volunteer post office is just established in which an express office will soon attach itself. A provision store will soon follow, the groceries, then dry goods, then a hotel, etc. until within 10 years the *tourist* of the continent will be whirled up to these diggings over a longer but far easier road winding around the mountaintops rather than passing over them and will sip chocolate and read his NY paper—not yet 5 days old . . . in utter unconsciousness that this region was wrested from the elk and mountain sheep so recently as 1859.

The overlap of activities was not in neat stages. Isabella Bird noted that tourists from the East were "trooping into Denver" and that the surveying parties were "coming down from the mountains." Bird, an Englishwoman, traveled in the Colorado Rockies in 1873, the year the Hayden Expedition was exploring and mapping the mountains. Her account was published as *A Lady's Life in the Rocky Mountains*. She visited Denver, saw mines, toured the spa hotels at Manitou, saw the sights at the Garden of the Gods, climbed Long's Peak, and sojourned in Estes Park, her "grand, solitary, uplifted, sublime, remote, beast-haunted lair." She found the wondrous scenery she sought. Comparing the Rockies with the Alps, she

found the Alps wanting. Her response echoed Zebulon Pike: "The scenery up here is glorious, combining sublimity and beauty. . . . I cannot by any words give you an idea of scenery so different from any that you or I have ever seen . . . this scenery satisfies my soul."

By the late 1870s, Henry Williams's *Pacific Tourist* recommended two weeks in the Colorado mountains. Here the splendors of scenery and the "mining attractions" merged. "We cannot possibly describe the attractions of these resorts. They are at once terrible, overpowering, lovely and full of indescribable majesty." Of this already blighted landscape, Helen Hunt Jackson predicted, "Perhaps 300 years hence the steep sides of the Georgetown canyon will be covered again with balsams and pines; the pinks, daisies, and vetches will carpet the ground as the pink heath does at Gastein; the mill wheels will stand still, the mines will be empty; and pilgrims will seek the heights as they seek Gastein's, not because they hold silver and gold, but because they are gracious and beautiful and health giving."

What brought, and continues to bring, visitors was the combination of the spectacular realities and the hyperbole of boosters. Boosterism and touristic hype are integral to the experience, where creating places and promotional imagery merges with reality. Colorado is a "wonderland," a "vacationland," the "nation's playground," the "Top of the Nation," "Ski Country USA," "at its best," "everywhere you look," and it "has it all." Here the past, as well as the landscape, is always colorful. The state's greatest attractions have been scenic, a great scenery sale, with thousands cashing in on the view. The nineteenth century's scenic admiration was based on European conventions for the picturesque, an emerging American idealization of the nation's wonders, and a landscape conforming to pictorial conventions. Thus photographs, engravings, and postcards reinforced expectations and confirmed the choice of destinations.

Access to sights as well as minerals was a major impetus for railroad and road growth into the mountains. Oftentimes, the routes themselves became attractions. Major Shadrach K. Hooper coined the Denver and Rio Grande Railroad's slogan "Scenic Line of the World," which boasted that the scenery changed every mile and that there was "scarcely a mile of tame scenery" along its mountain routes. From the mobile security of the railroad car, the *Pacific Tourist* described the leisurely pleasure of Pullman-car life. Looking out to the prairies, "where the eye sees but wildness and even desolation, then looking back upon this

**HARRY H. BUCKWALTER**
Passengers on Colorado Midland Excursion Train, June 15, 1901. *Courtesy of the Colorado Historical Society.*

long aisle or avenue," the tourist saw "civilization of comfort and luxury. How sharp the contrast." The view from the railroad path accentuated vistas and the precipitous character of Colorado topography; and with an engineer at the throttle, one could revel in the experience. Richard Harding Davis, an 1892 visitor from New York, warned that when riding the railroad through the Colorado mountains, "one commits a sin if he does not sit day and night by the car window."

In 1876, the year of statehood, the Atchison, Topeka and Santa Fe Railroad timetable promoted a compendium of Colorado's appeal and attraction: "The New and Direct Route From Kansas City and Atchison to Pueblo, Colorado Springs, Manitou, Denver, Canon City, Cuchara, Trinidad and all Points in Colorado. Six Hundred Miles of River, Plain and Mountain Scenery." It advertised the routes to the San Juan gold and silver mines along a route also called the Land Hunter's, Gold Hunter's, and Buffalo Hunter's Road. "Up the Great Valley of the Great Arkansas!" the brochure exhorted. "Lovely Scenery, Sublime Views, Recreation, Health, Pure Mountain Air, Trout, Game, Antelope, Buffalo, Elk, the Foot Hills, Mountains and Canons. The Shortest Route to Colorado Springs, Pike's Peak, Manitou, the Invalid's Heaven."

ROBERT DAWSON  Hot Springs, Glenwood Springs, 1985. From the Water in the West Project. *Courtesy of the artist.*

The railroad became much like a modern theme-park attraction through the mining and mountain landscape. Thousands visited the Georgetown Loop. In 1906 the Argentine Central ran a spur almost to the summit of Mount McClennan, a 4,500-foot climb in sixteen miles, on a narrow-gage traction railway. By the time of the Cripple Creek rush, tourism was well practiced in piggybacking its own bonanza on the mineral discoveries. As Duane Smith has noted, "The railroads came quickly, in fact they brought the tourist to see sights almost before the sights became seeable." New technologies described and mythologized places just as rapidly. One of the earliest motion pictures was Edison Film's *The Cripple Creek Barroom* of 1898.

Lines acquired identities. The D&RG promoted its Royal Gorge Line; the Silverton Railroad emphasized its Rainbow Route; the DB&W publicized the Switzerland Trail. The lines organized excursion trains into the mountains. Wildflower excursions took thousands to mountain meadows, and engineers stopped wherever wildflowers were blooming. In summer, people would take trains just to get to snow. In the 1880s, "fish" trains would stop

**LEE FRIEDLANDER**  Colorado Springs, 1972. Reproduced from: *The Great West: Real/Ideal. Courtesy of the artist.*

to drop off and pick up fishermen along the Platte Canyon. Later there were ski trains to Winter Park.

The image makers of the routes were the handmaidens of this burgeoning activity. William Henry Jackson began work for the Denver and Rio Grande in 1881. One of his most reproduced images was of a train passing through Toltec Gorge, seemingly hanging from the cliff face, for at one point a six-foot rock ledge was blasted one thousand feet over the water. The sixty-four mile Cumbres and Toltec Railroad, from Antonito, Colorado, to Chama, New Mexico, was built in 1879-80. Going through Cumbres Pass (Spanish for "summits") in the San Juan Mountains at 10,015 feet, it was the highest railroad pass in the nation. The line served passengers until 1951 and freight until 1968, only to be resurrected as a scenic railroad in 1971.

In 1960 the D&RG still had the Silverton Railroad, the ski train to Winter Park (which is again operating), the Vista Dome California Zepher, and the Denver-Glenwood Springs route going through Royal Gorge. Now, after the demise of the railroad as a viable means of passenger travel, tracks are abandoned everywhere, and the railroad station, formerly the proud, busy, focal point of a community, is adapted to another use, often as a restaurant with a railroad theme.

Roads allow freer movement, and scenic highways superseded the railroad excursion as high-country roads made the inaccessible open to the masses. Popular early scenic drives included High Line Drive to Cañon City, Big Thompson Canyon Road to Estes Park, and drives linking the elements of the Denver Mountain Park System. A popular loop from Denver passed through the stone gates of the Lariat Trail Scenic Mountain Drive at Golden, up Lookout Mountain, to Genesee Park, Bergen Park, Bear Creek Canyon, and Morrison, and then back to Denver. The Pike's Peak toll road opened in 1916; in the 1920s, Red Mountain Pass Wagon Road became the first section of the Million Dollar Highway between Ouray and Silverton; and Trail Ridge Road opened in Rocky Mountain National Park. All of these routes were designed with overlooks and wayside areas and were engineered for pleasure driving. In 1989, five Colorado roads were designated as scenic and historic byways under a Forest Service and Bureau of Land Management program: the San Juan Skyway through Durango, Silverton, Ouray, and Telluride; the Alpine Loop through Lake

**RUSSELL LEE** Tourists at Cliff Dwellings, Mesa Verde National Park, August 1939. *Courtesy of the Library of Congress, Farm Security Administration Collection.*

**TOD PAPAGEORGE**  Mesa Verde, ca. 1971. *Courtesy of the artist.*

City, Ouray, and Silverton; the Scenic Highway of Legends over Cuchara Pass; the Gold Belt Tour between Cañon City and Cripple Creek; and the Peak-to-Peak Highway between Estes Park and Black Hawk. Any "scenic" designation on a highway map will send thousands along a route.

Travelers need to be housed and fed. When Isabella Bird left towns, she found lodging in private homes, but hotels soon followed and were often the most glorious and palatial structures in the towns. Accommodations became attractions: hotels like the La Veta at Gunnison, the Grand at Silverton, the Beaumont at Ouray, Teller House at Central City, the Antlers in Colorado Springs, Hotel Colorado in Glenwood Springs, the Clarendon of Leadville, Beebee House in Manitou, Mount Princeton at Buena Vista, and the Ramona Hotel at Cascade, a Byzantine building with three floors of porches wrapping around its tower. The spas and resorts of Colorado Springs and Manitou were especially popular. The grandest, the Broadmoor in Colorado Springs, continues to develop to this day as a complete tourist village.

Whatever the glories of the great resorts and the splendors of their accommodations, for most visitors today the ubiquitous motel and roadside facilities and attractions set the tone, character, and standard of much of the travel experience. The highway strip linking road to town is a national landscape type. It has evolved to care for the needs of the vehicle—gas stations, car washes, auto showrooms—and to cater to the desires of the traveler—motels and drive-in functions of all types, theaters, banks, restaurants, supermarkets, drugstores. Businesses cluster in small pockets and in great malled enclosures. All are identified by abundant and accessible parking. With the exception of local businesses, the Colorado roadside is largely the product of national trends: common highway engineering criteria, franchising, and the benefits of media advertising. A road such as Colorado Springs Academy Boulevard, ringing the city to the east, is a compendium of 1980s roadside attractions with malls, mini-malls, franchises, and drive-ins of every conceivable enterprise, all in bright colors and bright lights.

One end of Durango's Main Avenue has the Durango-Silverton train station, tourist shops, a McDonald's in a railroad theme, and an outlet store for Ralph Lauren, the quintessential urban easterner who has managed to market a modern image of western style.

For several blocks, the tourist world is dominant, but as the street continues it is shared by the local inhabitants—Woolworth's is the dividing line. Across the state there is now a "cappucino curve." One can map where cappucino is served as an indicator of the spread of an urban, trendy culture.

Towns have dual identities. In heavily touristed communities, the balancing of old and new, of different cultures, is problematic. In a place like Steamboat Springs, old ranching and farming communities are mixed with skiers, recreationalists, and tourists. The traditional *land* recreation activities—hunting and fishing—don't change the landscape. The skilled outdoorsperson learns to read the land and then use it. New activities, skiing most dramatically, change the land and import a culture that stays long after the visitors have departed. For half the year, ski slopes sit like a great reservoir, where boats are beached anticipating the rising waters. So too skiers anticipate the falling snow and rising base levels.

Scenery was not the only attraction; salubrity was another. From its earliest days, the state has been a health mecca. Luke Tierney's 1859 guide to the gold country noted, "Some old mountaineers who have traded among the Indians here for many years, assert that they have never known a man to die in this region from any disease contracted here." Isabella Bird said that "nine out of ten settlers" she met were cured invalids and called Colorado "the most remarkable sanatorium in the world" with "health in every breath of air." The high-altitude, cool, dry air was recommended for those with tuberculosis (consumption), asthma, and respiratory diseases. It has often been noted that from 1880 to 1890, perhaps one-third (boosters said one-half) of the settlers were health seekers. The salubrious climate was supplemented by hot mineral springs, spas, radium tunnels, and sanatorium communities. An 1881 Denver and Rio Grande guide's three-part title, *Health, Wealth, and Pleasure in Colorado and New Mexico,* is a succinct summary of what was being sought by visitors and potential settlers. Even then, the virtues of escape were promulgated. "In the vastness of the mountains "elbow room" that they afford, in which to escape for a time from the fierce struggles and intense mental strain of modern civilization lies their chief attraction to many travellers."

Tourist attractions are created from natural features and manufactured elements; acting as magnets that attract and repel, they are finely attuned to their audiences. Early

adventurers discovered places, which were then codified by guidebooks as points of interest and organized as a touristic itinerary. Attractions oscillate between the authentic and the ersatz. There has been the odd continual development of attractions where none seem necessary: miniature golf courses, wax museums, artificial "genuine" ghost towns, the fake cliff dwellings at Manitou Springs, the night lighting of waterfalls. The paradoxes abound. There is clearly a genuine appreciation of natural wonders and beauty, which exist side by side with the artificial, even the bizarre. Heading out of Colorado Springs, the traveler reads the sign pointing to Pike's Peak in one direction—the North Pole in the other!

People create landscape symbols and enact landscape rituals. Much of the touristic landscape is composed of sites and attractions that enshrine and reenact the encounters of people and place, from explorers and traders to soldiers and settlers. The history is often presented in confused and debased form, but it is present nonetheless. In the process, history's tragedies and processes also assume their place. In the early portion of the century, the indigenous cultures, the displaced Indian tribes of the West, became depersonalized as "attractions." Wildlands, territory the Anglo-American settlers imagined as "uninhabited," also became an attraction and preserve, not a culture's habitat.

What's the difference between touristic and tacky? When is history abused and trivialized beyond repair? The tacky is often found on the Main Streets, where often there is a passage through a purgatory of T-shirt and souvenir shops en route to a destination. Tourist towns and national parks have similar characters: visitors crowd the "centers" and designated touring sites, but just off the marked and signed trails lies little-frequented backcountry. Touristic landscapes can become Disneyfied; a place is created as an ersatz phenomenon, or actual places are perceived as artificial attractions. With Disneyland as a cultural reference, almost anyplace has the capability of becoming a "-land." The suffix can transform places into attractions, where they may join a pantheon of places, implying enjoyment and a simple and sanitized history. These new "lands" are not to be sold or worked but visited. Thus old mining towns have a theme-park quality; they become Miningland, Gold Rushland, or Silverland, and National Park sites become National Parkland.

Some sights and sites have become touristic and cultural icons, coded symbols of the state. These icons are trivialized on calendars, T-shirts, plates, matches, or spoons, but they

**ROGER MERTIN**  Goldpanning at the Old Timer, Near Idaho Springs, 1978. From the portfolio "From This Land." *Courtesy of the Colorado Historical Society and the artist.*

**WILLIAM HENRY JACKSON**   Mountain of the Holy Cross in the Great National Range, 1873. *Courtesy of the United States Geological Survey, Denver.*

"No man we talked with had ever seen the Mountain of the Holy Cross. But everyone knew that somewhere in the far reaches of the western highlands such a wonder might exist. Hadn't a certain hunter once caught a glimpse of it — only to have it vanish as he approached? Didn't a wrinkled Indian here and there narrow his eyes and slowly nod his head when questioned? Wasn't this man's grandfather, and that man's uncle, and old so-and-so's brother the first white man ever to lay eyes on the Holy Cross — many, many, many years ago? It was a beautiful legend, and they nursed it carefully. " From *Time Exposure: The Autobiography of William Henry Jackson.*

carry significant symbolic value and meaning. The simplest guide to the changing iconography of the state may be found in postcard collections. In the nineteenth century, the primary sites included the Garden of the Gods, the Mountain of the Holy Cross, and Pike's Peak. A 1914 D&RG album included Manitou—"the Paradise of the Colorado Tourist"—Skyline Drive, Royal Gorge, Mount Princeton, Toltec Gorge, and the Canyon on the Grand River. From 1906 until 1931, travelers to Denver were greeted by a great Welcome Arch lit by sixteen hundred lightbulbs. It was originally intended to be torn down when the new station was built, but for a time it was a beloved symbol. It was eventually demolished as "unsafe"; at the time there was little money or will to repair it, and the feeling was that its removal would improve the city's image. New sites and sights have been added to the state's iconography: Maroon Bells, the mill at Crystal, the Air Force Academy, Denver's skyline, Mesa Verde, the National Center for Atmospheric Research, and always mountains, from peaks to ranges.

However, if we take T-shirts and airport shops as guides, it is the winter landscape that is now fundamental to the state's iconography and imagery. In recent years the salubrious climate has been superseded by an attraction of winter—snow. Colorado was one of the inventors and incubators of American ski culture, the discovery of winter and of mountain sports. Colorado skiing changed the age-old patterns of transhumance: following one's animals in search of fresh food and fodder; going to the mountains only in spring and summer and abandoning them in winter. Snow, which represented hardship and was a burden to be removed, became a bellwether of progress, and each fall its arrival is eagerly anticipated.

Isabella Bird found "snow patches, snow slashes, snow abysses, snow forlorn and soiled looking, snow pure and dazzling, snow glistening above the purple robe of pine worn by all the mountains." The mountain snows now beckon and call skiers seeking the perfect conditions. Champagne powder became a new landscape ideal and was added to the touristic vocabulary. Nineteenth-century Scandinavian immigrants brought skis, then called snowshoes, to the United States. However, for skiing to succeed as a form of mass recreation, a new transportation technology was needed to make mountain slopes accessible. The first rope tow was built in 1937 on a Pike's Peak hillside by the Silver Springs Ski

Club of Colorado Springs. This was followed by T-bars, V-bars, and Poma lifts. In 1939 the first chair lift was constructed at Gunnison Pioneer Ski Area, using a modified ore tramway system. Chairs for two, three, and four persons followed, and with each new development, speed and capacity increased. Gondolas were first used at Vail in 1962, taking the name of Venetian boats.

Camp Hale, near Leadville, the home of the army's Tenth Mountain Division of ski troopers, introduced the Colorado mountains to these soldiers. Many returned: Vail, Breckenridge, and Arapahoe Basin were founded by veterans as, after World War II, skiing began its dramatic boom. Aspen, Breckenridge, Crested Butte, and Telluride, all former mining towns, were resurrected by skiing. Snow was "white gold," and a new rush was on, only now with a double meaning of riches and thrills. Mountain development, lift technology, improved equipment, highway access, and the marriage of automobile and the ski-rack all contributed. Skiing is only the most dramatic example of a landscape designed for recreation. Each recreational type spawns a recreation culture of users and advocates. There are skiers, rafters, white-water enthusiasts, bikers, hikers, hunters, balloonists, and birders. All of these groups, making their demands on the landscape, are sensitive to different aspects of landscape change.

Mountains, once idealized from afar, are now shaped and carved, made accessible for winter sports. Skiers have their own landscape criteria. They are connoisseurs of snow conditions, terrain, challenge, and excitement. Although the scenic is still important, the mountain experience is no longer primarily visual or symbolic but rather is visceral, interactive.

**ROBERT DAWSON**
Headwaters of the Colorado River, Rocky Mountain National Park, 1986. From the Water in the West Project. *Courtesy of the artist.*

**S. W. MATTESON**  Ute Indian Settlement, ca. 1897. *Courtesy of the Denver Public Library Western History Department.*

# SETTLEMENT

*"It is like a city of fresh card-board."*

To SETTLE IS TO COME to terms with something. When an argument is settled, there is an agreement—so too with settlement on the land. Settlements represent an attempt to reach an agreement, a satisfactory accord, between people and place. Wallace Stegner has written that western towns represent a balance between mobility and stability. There is a paradox, perhaps irreconcilable: how can one "settle down" if movement, mobility, restlessness, and change are endemic in the American cultural character? In the West this dichotomy is accentuated by the scale and character of the landscape. In vast spaces, towns stand out, and the characteristics of a settled community contrast dramatically with the surroundings, which are often of a more wild nature. An 1876 AT&SF land advertisement—extolling the virtues of the Colorado climate, soil, timber, stock, and coal—noted, with no regard for the indigenous culture, "You are not compelled to go into a wilderness, but into a country settling rapidly, where you will find immediately good society, good newspapers, good churches and good markets, with good schools and free education for your children." This struggle to build a "good" society was built into the character and design of settlements.

Colorado communities include basic American settlement types such as the Main Street town, the product of nineteenth-century commercial development, and also the automotive suburb of the postwar generation. But there are other types more distinctive to the state: cliff dwellings, camps, ghost towns, colony towns, and ski resorts. These settlements are not exclusive to Colorado, but it is here that they have developed in dramatic fashion.

For the native inhabitants of Colorado, the land was home, provider, and spiritual presence. Territories were loosely defined into tribal hunting grounds, but the concept of alienabil-

## CLIFF DWELLING

**WILLIAM HENRY JACKSON**  Street in Lower Creede, early 1880's. *Courtesy of the Colorado Historical Society.*

**JOSEPH BEVIER STURTEVANT**  Topping Avenue, Chautauqua, Boulder, ca. 1899. *Courtesy of the Carnegie Branch Library for Local History, Boulder Historical Society Collection.*

ity—that the land was property to be purchased or bartered—was foreign to their cultures. The Arapahos and the Cheyennes claimed much of the territory between the South Platte and the Arkansas rivers, the Kiowas and the Comanches south of the Arkansas, and the Utes, the Shoshonis, and the Apaches the lands of the mountains and high plateaus.

These tribes were all seminomadic and nonagrarian. Their dwellings and settlements conformed to a way of life in which villages moved to follow game and to seek better quarters. Sites reflected the requirements of a people living off the land. They needed water, security, proximity to food supplies, and grazing land for horses introduced after the Spanish conquests. Plains Indians dwelled in tipis made of buffalo hide. Francis Parkman described one plains settlement in "an admirable position. The stream with its thick growth of trees enclosed on three sides by a wide green meadow, where about forty Dahcotah [sic] lodges were pitched in a circle, and beyond them a few lodges of the friendly Shiennes [sic]." Frémont visited a Cheyenne and Arapaho village of 125 lodges, which "were disposed in a scattering manner on both sides of a broad irregular street, about one hundred and fifty feet wide and running along the river." Villages could have several hundred lodges, each housing up to ten persons, yet whole villages were easily dismantled and moved. The evidence of the indigenous culture of these nomadic hunters was easily erased, and their traces are almost invisible. However, their presence underlies the landscape. The Indians hunted buffalo before ranching domesticated livestock, they were the first farmers and irrigators, they built the first towns and the first paths across the land. The spirits of their dead still reside here. Subtle remnants of their influence abound in modern communities, which often chose traditional Indian locales for townsites, farmsites, and routes.

Modern visitors are not prepared for the cliff dwellings of Mesa Verde. The dwellings are not where buildings are supposed to be but are located in perches that are typically reserved for other species, the nests of birds or the hives of wasps. These balcony cliff sites are Colorado's grandest porches. The stone curves and arches over ledges where dwellings are built, framing the sky in great arcs, like a giant cinemascope screen. Within these spaces the primary elements of landscape have a visceral presence, with no dividing lines between dwelling, village, and nature. Dwelling is in close contact with the earth, and in the subterranean kiva there is an embracing ceremonial presence.

There are many lessons here in the complete integration of site and structure, building and land, earth and sky. Marks are clear and long-lived in the dry southwestern landscape. The cliff faces have handholds and footholds, giving the rock human dimension and measure, much like the carefully modulated proportions of the classical orders of ancient Greece. Why did the Indians dwell in this precipitous situation? Why did they leave? Early theories suggested that the Anasazi chose these sites for security, but that is no longer thought valid. They may have exhausted and overused the landscape resources of arable land and available wood for fuel and building. The climate may have changed; it became colder, harsher. These reasons are all rational, but what of the idea of challenge? The cliff dwellings recall monastic sites in the Egyptian and Judean deserts—places of testing one's faith. William Henry Jackson found the houses "perched away in a crevice like a swallow's or bat's nest, it was, a marvel and a puzzle." They remain so and continue to challenge us to rethink our notions of dwelling.

## PLAZA

Hispanic frontier settlements were called *plazas*. The plaza was also the central village gathering space, since physical and social spaces were coexistent. Plazas were built of *chorreras*, lines of adobe houses with common walls for defense and identity, with a loose grid of streets converging on the center, the place of celebration, meeting, and market. Surrounding the village, farmland was apportioned by need, with grazing lands and territory in the mountains held in common, territory essential for fuels, building materials, wildlife, and gathering. Private farmlands were in long strips measured in varas, equal to about thirty-three inches. Strips measured from twenty to five hundred varas wide and from a few hundred feet to miles long. These linear strips, ribbon farms, were then cut into stripes of garden, pasture, and field, often leading into hillsides; therefore everyone would be assured of a variety of lands, from fertile bottomlands to wooded slopes.

The Plaza de San Luis was founded in 1851, and within three years the communities of San Acacio, San Pedro, San Francisco, and Chama were established. All were part of the Sangre de Cristo grant, issued in 1843 but still recognized by the United States after this part of Mexico was ceded. The plazas represent the surviving remnant of a traditional agricultural landscape, a compact village, with its central focus and communal endeavors in land

ownership, distribution, and irrigation. The settled and communal character of these enclosed adobe agricultural communities contrasted strongly with the transient nature of mining camps, which were founded at the same time. As Anglo communities encroached on these areas, many plazas were abandoned. The legal basis for ownership, Mexican land grants, or *la Merced*, were contested, and traditional village life declined.

Urban areas on the western frontier had a different pace and pattern from eastern communities. These rapidly growing towns have been termed "instant cities," in which the processes of city building were compressed into a brief span of time. The gestation period for these towns was brief, and they often emerged not quite fully formed. In basic pattern and form, they were certainly recognizable as American communities, but their often isolated locations accentuated their qualities.

Bayard Taylor said that the mining camps in Gregory and Nevada Gulch had a "curious, rickety, temporary air, with their buildings standing as if on one leg, their big signs and little accommodations, the irregular, wandering uneven street, and the bald, scarred, and pitted mountains on either side." He added, "It is a place . . . where everything is old, grotesque, unusual." Mining camps often began around streams; the resource was primary, the work was right there, and accommodation was secondary. The goal was to stake and work one's claim, and only later did anything beyond the bare essentials of community life develop.

The camps came first. Many developed no further, but others evolved past the primary stage into towns. Helen Hunt Jackson said of Leadville's transformation: "In six months a tract of dense spruce forest had converted into a bustling village. To be sure, the upturned roots and freshly hacked stumps of many of the spruce trees are still in the streets. . . . Some of the cabins seem to burrow in the ground; others are set up on posts, like roofed bedsteads—tents, wigwams of boughs . . . cabins wedged in between stumps, cabins built on stumps; cabins with chimneys made of flower pots or bits of stove pipe . . . cabins half-roofed; cabins with sail cloth roofed, cabins with no roofs at all—this represents the architecture of Leadville homes." By the end of that year, 1879, Leadville had 31 restaurants, 17 barber shops, 51 groceries, 4 banks, 120 saloons, an opera house, a red-light district, schools, lodges, social clubs, gaslights, telephones, denuded hillsides, and the ever present smoke from smelters.

## CAMP

**Photographer Unidentified** City Park Auto Camp, Denver, ca. 1920. *Courtesy of the Denver Public Library Western History Department.*

Tents gave way to log cabins, then a sawmill was built, followed by frame houses, shops, sidewalks. Work would start on a school and a church, and brick and stone would make their appearance as building materials. Along Main Street would be found the necessities of a general store, a livery stable, a saloon, and a boarding house. To these were soon added a newspaper, more saloons, and as a sign of cultural aspiration, a theater. Towns founded by streams crept up hillsides. The civilizing agents—schools, churches, homes of leading families—found their position above Main Street, whereas saloons and gambling were to be found below. At first, mining camps were predominantly communities of young men, but soon husbands were joined by wives and children. Wives brought a domesticity, the desire to "settle down" and create a community. The dialectic of wilderness and civilization was enacted out in social codes as well—the school and the church versus the saloon.

The process reoccurred with each rush. Creede was home to ten thousand persons in 1889 during its silver boom. By then, towns were immediately linked by telegraph, and after a ten-mile spur was built from Wagon Wheel Gap, one could take the D&RG Railroad directly to Creede. Richard Harding Davis visited Creede when it was just twenty months old. In *The West from a Car Window* (the window was that of a railroad car), he wrote that the area of Willow Gulch was "tramped into mud and covered with hundreds of little pine boxes of houses of log-cabins, and the simple quadrangle of four planks which mark a building site." He added: "In front of you is a village of fresh pine. There is not a brick, a painted front, nor an awning in the whole town. It is like a city of fresh card-board. . . . It is more like a circus tent, which has sprung up overnight and which may be removed on the morrow than a town." Yet even this encampment of a town was lit with electric lights.

The mining camps were remote and rugged. They were nascent cities of youthful populations. Working against their stability were the centrifugal energies of a transient population, speculative economics, and the lure of opportunity down the road. The tension between stability and mobility, between putting down roots and pulling up stakes, was always present. Some settlements demonstrated a remarkable tenaciousness and soon stabilized, but many dissolved.

Mining camps had a makeshift, do-it-yourself quality, a seat-of-the-pants inventiveness. These qualities were shared by many Colorado settlements. Camps were temporary

**JACK ALLISON** Great Western Sugar Company's Colony for Sugar Beet Workers, Hudson, September 1938. *Courtesy of the Library of Congress, Farm Security Administration Collection.* "A colony of 20 adobe houses built by the inhabitants with materials supplied by the company. 13 of the houses are used, 7 being unfit for habitation. Aproximately 50 people live in the 13 houses. Being in the limits of an incorporated town (Hudson) there is a water system. However, there is only one outlet (an outdoor spigot) for this whole colony. There is no electricity, gas, or sewage disposal." Caption from print.

communities with uncertain futures; they were gatherings of strangers brought together by a common desire. Camps accentuated the poles of rootedness and mobility in the American character, as people followed the path of gold, beavers, sun, or snow. Camps represented a combination of promise and insecurity and demonstrated that community need not always be synonymous with a fixed locale or longevity. Camps were common in the Colorado experience: the rendezvous sites of mountain men, cattle camps, health camps, auto camps, CCC work camps, relocation camps, military camps, Chautauqua, and summer festivals.

The mining camp may be the prototype, but it had precursors. Noting that the camp of the mountain hunter was "invariably made in a picturesque locality," George Ruxton wrote: "Like the Indian, the white hunter has ever an eye for the beautiful. The broken ground of the mountains, with their numerous tumbling and babbling rivulets, and groves and thickets of shrubs and timber always afford shelter from the boisterous winds of winter, and abundance of fuel and water. Facing the rising sun the hunter invariably erects his shanty, with a wall of precipitous rock in the rear to defend it from the gusts which often sweep down the gorges of the mountains." Every summer between 1825 and 1840, at the height of the beaver trade, a rendezvous of mountain men was held at Brown's hole in northwestern Colorado as these individuals gathered for trading and camaraderie.

Even in their temporary camps, explorers sought satisfying locations. The Long Expedition chose a site near Denver. "Our camp is beautifully situated on the bank of the river, which is here about 100 yards wide—our tents pitched in a grove of cottonwood trees, that shade us from the scorching rays of the sun." At the mouth of the Platte canyon, they encamped "on a small plain of the river bottom which afforded good feed" for their horses. With "a number of large cottonwood trees" to shade them "from the rays of the sun" and "good cool water from the river," they "set the flag on the hill and a sentinel to look out."

Cattle camps were mobile caravans of chuck wagons and riders. A wagon was associated with ten to twenty riders, each with a string of six to ten horses. Each wagon had a "cavyyard," a corruption of the Spanish *caballado*, of sixty to one hundred animals. Camp moved six to ten miles per day. At camp, cattle were "cut" into herds by brands, and during spring roundup, branding took place.

**JOE MCCLELLAND** Granada Japanese Relocation Center, Amache, 1943. Photograph made for the War Relocation Authority. *Courtesy of the Denver Public Library Western History Department.*

**WILLIAM HENRY JACKSON** The Old Town, Pueblo, ca. 1900 (both pages). *Courtesy of the Denver Public Library Western History Department and the Library of Congress.*

Health camps were created for "lungers," those suffering from tuberculosis. The Colorado *Souvenir Book for the International Congress on Tuberculosis* of 1908 had articles on "How Does Colorado Climate Influence Tuberculosis?" and also "Touring, Camping, Hunting, and Fishing in Colorado." An illustration of tipis was captioned "The First Campers"! In Edgewater were the camps of the Jewish Consumptives Relief Society, the Evangelical Lutheran Sanitarium for Consumptives, the Swedish American Sanatorium, and the Y.M.C.A. Health Farm. In Colorado Springs were found the Nordach Ranch Sanatorium and the Modern Woodmen of America Sanatorium. Mini-cities of tents and small cabins, they were the predecessors of the respiratory medical centers in Denver that today serve thousands.

Before the institutionalization of the motel, the autocamp was the preferred stop of travelers. Denver's City Park was home to the city's first municipal autocamp in 1915. The city soon boasted of having "the Manhattan of auto-camps," a mini-city at the Overland Motor Park, which opened in 1920. By the summer of 1923, almost sixty thousand "tin can tourists" camped there. Autocamps were instant, brief communities, where travelers gathered for a week or two. Overland Park included a two-story clubhouse with grocery, meat market, lunch counter, billiard room, barbershop, laundry, and dance floor. The one thousand sites filled 160 acres along the South Platte River. Along the P.P.O.O. Highway were camp groves with water, electric lights, and comfort stations at Burlington, Stratton, Seibert, Simla, and other locales. Salida had three camps with a one-thousand-car capacity. These early autocamps were the forerunners of the hundreds of campgrounds and RV parks now found across the state.

Other camps—the hobo camps common along railroads in the depression, the camps of migrant farmworkers, or the tent colonies of striking miners—didn't come with the amenities. World War II brought two disparate camps to the state. In the mountains near Leadville was Camp Hale, the training ground of the Tenth Mountain Division. Near Granada was Amache, the more infamous site of the Japanese-American Relocation Camp, where ten thousand persons were interned. It was named for Amache, the daughter of Chief One-Eye, a Southern Cheyenne killed at the Sand Creek Massacre.

## BOOMTOWN, GHOST TOWN

Mining camps were Colorado's first boomtowns. Even the most successful went through successive stages of boom and bust. Most were short-lived settlements, but some grew into supply towns or smelting centers or became revitalized with new mineral discoveries. A few would boom again in a new guise, as tourism or skiing capitalized on their locale, lore, and building stock. One can now see places in different stages of this process, from the lone cafe of Saints John, a ghost town above Montezuma, to Aspen or Breckenridge.

Plat maps were drawn up before settlement, as enticements for settlers. Their vacant spaces were egalitarian, available to all those who had the cash. They were like great urban coloring books waiting for lines to be filled in. Wide streets were practical for wagons and animals, but perhaps more important, they were a symbol of ambition and expectation—as were plats for parks and squares. In the transient, new atmosphere of the boomtown, the wilderness was conquered, and urban devices were rapidly adopted: street lighting, the newest transportation, telephones. The works of early artists and photographers accentuate this contrast, showing newly settled communities with wild nature knocking at their doors.

Boomtowns exaggerate certain characteristics. Boosters help foster an atmosphere of future greatness. Boomtowns are boastful, with big signs, gaudy displays, and fancy facades. They worry little about what's out back or what's coming tomorrow. They are speculative by nature, gambling on the future. The boomtown quality persists a century later. The Front Range communities boomed after World War II, with settlers attracted by the state's climate, scenery, and recreation and the money brought in by the federal government, the military, high-tech industry, and energy-related industries. In the early 1970s, Aurora was the fastest-growing midsize American city. Demand for new energy supplies boomed the towns of Craig, Meeker, Carbondale, Rangely, and Somerset. Exxon's Colony Project planned the new town of Battlement Mesa, seventeen miles from Parachute, but the project was dropped when oil prices declined. Bust was immediate.

It is hard on towns to expand and contract. Some blow up like balloons and burst while others slowly deflate. Older communities often are initially enthusiastic about the prospect of booming, as a solution to their dilemmas, but booms bring their own problems: strained relationships between long-term and new residents; the difficulties of keeping up with the needs of rapid population growth. Sometimes disaster is the determinant. Poncha Springs,

**KENNETH HELPHAND**  Deserted Street, Uravan, 1989. *Courtesy of the author.*

**KENNETH HELPHAND**  Uravan, 1989. *Courtesy of the author.*

**ELLEN MANCHESTER**  Abandoned House, Uravan, 1985. *Courtesy of the photographer.*

**ELLEN MANCHESTER**  Abandoned House, Uravan, 1985. *Courtesy of the photographer.* A modern ghost town, Uravan was settled in 1936 to house workers during the uranium and vanadium boom on the Western Slope. It was abandoned in 1984 after the mine closed and widespread radioactive contamination was discovered throughout the town and in the workers' homes.

forty miles southeast of Bonanza, was destroyed by fire. Maggie Brown wrote home to her relatives in Virginia, "The town was about dead and now I suppose it will die." One early settler said, "Most of the towns I was interested in played out."

Those that die become ghost towns. Ghost towns accentuate the transitory nature of settlement and experience. In these weathered, windowless, doorless, rusting places, nature is patiently reclaiming its position. Ghost towns are haunting not haunted; their flaking, peeling, and rotting buildings are signs that mark the passage of time. The abandoned Anasazi cliff dwellings were perhaps the first Colorado ghost towns, but ghost towns need not be aged. The boom-and-bust cycle operates on a very short frequency. In 1869 Yale Professor William Brewer wrote of the already abandoned Buckskin Joe diggings in South Park, "Such places have a fascination for me—the old sign boards in the streets, the roofless houses, the grasses growing in the old hearthstones and flowers nestling in the nooks of muddled walls or broken chimneys, the multitude of empty fruit cans, sardine boxes... lying in the streets, bits of old saddles, rusty prospecting pans, old shovels, the stamps from mills—in fact, all the varied implements of a city, rusting and rotting, neglected."

Ghost towns have become attractions. Crystal is a ghost town, but it is a being brought back from the dead by new residents. On a visit we saw photographers, dirt bikers, climbers, hikers, an Outward Bound group, ATVs, jeeps, and four-wheel-drive vehicles. Even habited towns can have a ghostly feeling. Zane Grey wrote of Yampa: "The main and only street was very wide and dusty, bordered by old board walks and vacant stores. It seemed a deserted street of a deserted village."

Uravan is a modern ghost town that reeks of failure and fear, not nostalgia. A product of the 1940s uranium boom (*Uravan* is a contraction of *uranium* and *vanadium*), the town now displays grass and sagebrush growing in the cracks of the asphalt streets that led to the mine. One can enter the open doors of houses in this ghost subdivision and see wallpaper peeling off the walls, and cracked linoleum; outside, picket fences mark unkempt yards. It feels like a scene from the apocalyptic post-nuclear bomb disaster films of the 1950s. You do not linger. Triangular radiation-warning signs are on all the mine fences.

## COLONY

The diverse desires of the American dream take many forms, represented by the two nineteenth-century Colorado extremes: the mining camps of the mountains and the agricultural colonies of the Colorado Piedmont. These two represent the magnetic poles of the American ethos. On the one hand, there is an extreme individualism, and on the other, a shared vision of community. Whereas boomtowns of minerals, energy, and business have been bent on quick riches, other communities harken back to America's colonial past and the idea of a covenant community, a settlement based on voluntary association. In Colorado these are the colony towns, communities of people bonded and banded together by a civilizing sense of mission, people united not by schemes of rapid riches but by utopian aspirations, typically wedded to an agricultural community. In reality, all communities have elements of both extremes—the desire for individual wealth and wish fulfillment and the satisfaction of being a group member.

Nathan Meeker had visions for Colorado. Meeker had lived at the Trumbell Phalanx, a Fourierist utopian community in Ohio, for three years and was the agricultural editor for the *New York Tribune* when he founded the Union Colony in New York City. At a meeting at the Cooper Institute, the criteria for a colony site were agreed upon: "We should seek 1st, healthfulness; 2nd, a varied and rich soil suited for grass; 3rd, coal or timber, if possible both; 4th, iron ore; 5th, the adaption to fruit; 6th water power; 7th beautiful scenery." The criteria clearly suggested both an agricultural and an industrial community and echoed those of expeditions of a century earlier. A group was sent west and purchased seventy thousand acres of railroad and government land on the Cache la Poudre River. In a promotional circular, Meeker extolled the virtues of the location with a booster's zeal: "A location which I have seen is well watered with streams and springs, there are beautiful pine groves, the soil is rich, the climate healthful, grass will keep stock the year round, coal and stone are plentiful, and a well travelled road runs through the property.... In addition, the Rocky Mountain scenery is the grandest, and the most enchanting in America. I have never seen a place which presents so many advantages and opportunities."

The agricultural colonists of Union Colony came together through community desire and pragmatic need. The colony tried to balance American individualism with community, but not communal living. There was private property but also collective land purchase, individual family dwellings but also concerted, cooperative labor for the vast tasks of irrigation and fencing. For the $155 membership fee, a colonist received a parcel of farmland and the right to buy a lot in town. The plan called for a village settlement divided into ten-acre lots subdivided into eight lots for building. Each family would then be given acreage adjoining the village. Meeker likened the scheme to the settlement of New England villages; the orientation was toward traditional family life. "The highest ambition for a family should be to have a comfortable, and if possible, elegant home, surrounded by orchards and ornamental grounds, on lands of its own," wrote Meeker. This emphasis on and pride in the domestic landscape were only slight exaggerations of the virtue seen in the home environment nationally.

Meeker explained his vision of the common good: In planting, in fruit growing and improving homes generally, the skill and experience of a few will be common to all, and much greater progress can be made than where each lives isolated. . . . I make the point that two important objectives will be gained by such a colony. First schools, refined society, and all the advantages of an old country will be secured in a few years . . . second, with free homesteads as a basis, with the sale of reserved lots for the general good, the greatly increased value of real estate will be for the benefit of all the people, not for schemers and speculators." Greeley was selected as the name for the town, and colonists went through the rites of settlement. Irrigation ditches were surveyed and dug, and the forty-five-mile-long fence was begun. By 1871 the first ditch was twenty-seven miles long, the population was fifteen hundred, and thousands of maple trees had been planted. The 1873 town plan and plat map show Lincoln Park, Island Grove Park, the Agricultural College, and the cemetery as open spaces; all remain today.

This was a landscape of rectitude, hard work, and virtues that harkened back to the Roman ideals of citizenship—to be a man of both town and country. David Hamer has called Greeley an "anti-frontier" town for its law-abiding nature and emphasis on family, home, and temperance. It was dramatically different from contemporary mining camps. Compare the difference between agricultural plat maps and the overlaid and overlapping claims of miners. Examine the regularity, order, and civility necessary to build, maintain, and manage complex irrigation systems versus the first-come, first-served culture of the gulch. Look at

**Photographer Unidentified**
Greeley, Looking Down 8th Street Toward the Depot, 1870. *Courtesy of the Denver Public Library Western History Department.*

the water nourishing irrigated crops versus the sluices and hydraulics washing away the soil. Viewing the town from the mountains, Meeker compared it to a garden. "The lots [are] beds, and the streets, paths, even though it is only 90 days since a plow turned the first furrow."

There were many experiments at community building. General Robert A. Cameron, a Union Colony member, was active in promoting the Chicago-Colorado Colony at Longmont, Fountain Colony at Colorado Springs, and Union Colony No. 2 at Fort Collins. Fort Amity was established in 1898 by the Salvation Army in southeastern Colorado, but it closed ten years later. There were Mormon colonies at Ephraim, Richfield, Sanford, and Manassa and a failed rural Jewish settlement at Cotopaxi. John Osgood built Redstone as an ideal mining town in 1902. O. T. Jackson founded Deerfield in 1910 as an African-American settlement. In 1920 it had seven hundred residents, only to be abandoned a decade later. All that remains is the poignancy of a road sign, a few dilapidated buildings, and overgrown foundations—a modest monument to high aspirations.

The "colony town" was also a marketable idea. Railroad companies took the concept and made it into a real estate scheme, promoting developments that were colonies in name only. The D&RG and associated companies helped develop Monument, Colorado Springs, Fountain, South Pueblo, Walsenburg, El Moro, La Veta, Alamosa, and Durango.

## MAIN STREET

The most common Colorado settlement is the Main Street town. The parts and pattern are clear: an almost square town divided into rectangular blocks; a commercial street down the center; courthouse, library, and schools clustered near the core; a city park; a penumbra of houses set within trees; industry at the edge; the highway leading into and out of town. In most locales, after only a brief walk or bike ride, you are in the country. There are connections to other places; a railroad line diagonally slices through the town, a city street becomes a highway strip, then a rural road.

These towns have dimensions. The simplest is a single street competing with the through roadway. Traffic barely slows down. A town's pride was once expressed by a paved main street, cement sidewalks, and blacktop or "oiled" roads leading into the countryside. Townscapes are landscapes. Main Street is both a divide and a front range. Taller, false-fronted buildings, peaked with roofs, slope back to side streets. Side streets are like small valleys, residential tributaries leading for only a few blocks off the main street, the commercial and civic center of the town. Main Street towns are distinct in their territories, especially their subdivisions of social class, and every town has its "other side of the tracks." The "Jungles" of Fort Collins, the home of German-Russian immigrant sugar beet workers, consisted of "one room-shacks with stove pipe chimneys sticking out of the flat roofs at crazy angles, their weathered sides hugging the dirty yards, sagging fences hemming in dirty children . . . all drab, colorless, with here and there a brightly painted house, a good fence, only added a sordidness to the scene," wrote Hope Williams Sykes.

Wallace Stegner feels that "the law of sparseness" has kept small towns from growing too big. "They are the places where the stickers stuck and perhaps were stuck in a forlorn rightness. They are at once lost and self-sufficient, scruffy and indispensable," with a "loneliness and vulnerability." Yet, as Robert Athearn has written, his small town in 1920 had twice-a-day railroad service, telegraph and telephones, a daily newspaper, and home mail. "It was not out of touch with the world." It also had hardware, clothing, furniture, jewelry, department, and dime stores, drugstores, hotels, banks, a library, a lodge, and theaters, all selling products and services available in larger communities.

Colorado's Main Street towns have distinctive regional qualities. On the treeless plains and the Western Slope, many towns, like Meeker, feel like oases. Mountain towns, such as Silverton, with just a single paved street, look like urban swatches, small city samples placed in a wilderness setting. Early photographs show how many towns slowly "filled in" the ambitious plats. There is often an open quasi-rural character, with vacant lots, an occasional horse, and houses set within large treed parcels. The "grain" is loose. In dozens of towns, tight orderly square blocks are now surrounded by the newer additions penetrated by curved and cul-de-sac streets.

There are odd Main Streets. Orchard is a real town along the South Platte. It was also the location for the filming of the television miniseries "Centennial," based on James Michener's novel. From a distance, Orchards buildings all appear genuine, but closer inspection reveals one side of the street to be brick structures and the other side to be the peeling false-fronts of a stage set. Is there a lesson here for the slipshod construction practices and instant communities along the Front Range?

Many towns are now experienced from the highway. Entering from these routes makes it difficult to decompress to the pace and tone of the town. Most visitors remain in a narrow band of motel and restaurant, rarely venturing, even a single block, beyond the main street. Entering a town from its "backside," where the hinterland and the edge merge, often demonstrates the traditional linkage between town and countryside, as farmsteads give way to houses and as farmyards become just yards. The more common entry is along commercial routes, where the street is a vehicular shopping mall. A few communities retain the traditional commercial Main Street, and others—including Grand Junction, Pueblo, and Aspen—have tried to recapture the Main Street spirit in restorations and revitalizations. The Boulder Mall is perhaps the most successful as a community center; on a grander scale, Denver's 16th Street Mall has catalyzed downtown activity.

In Walden, the Jackson County Courthouse crowns the hill. It sits Acropolis-like, up against the striking North Park landscape. It is a bold, civilizing statement. The other tall structure in town is the water tower. One is a symbol of social ideals, the other of reality and practicality. Equally civilizing in this harsh isolated setting are the modest signs of community life along Main Street: cafe, theater, bank, drugstore, gas station. There are no T-shirt shops.

## DENVER

We typically think of settlements in spatial terms, from small town to metropolis, but they also have a temporal measure. Denver was, in Gunther Barth's terminology, the quintessential "Instant City," populated by transplants, without tradition, adapting to a wild setting and living for the moment. Instant cities compressed the processes of development, improvising as they went along. Paradoxically, they tried to be more citylike, a factor of their isolation and impatience for growth. They were clean-slate landscapes, starting anew; they had little awareness and no respect for what was being erased and gave only modest forethought to ultimate consequences. These communities lived on a spatial and psychological edge. The spirit was speculative, a bet on urban survival. Success was not assured, and failure was common. The atmosphere was exhilarating but insecure. Still, conventional civic concerns asserted themselves, and transplanted institutions provided a stabilizing force to the city's centrifugal energy.

When Horace Greeley visited, he reported: "Of these rural cities, Auraria is by far more venerable—some of its structures being, I think, fully a year old if not more. Denver, on the other hand, can boast of no antiquity beyond September or October." He saw no floors in any buildings. Illustrations of Denver in 1859 show cabins, tipis, wagon roads going off into the distance, and mountains on the horizon. By midsummer the city had three thousand cabins. Samuel Bowles, comparing the changes in Denver from 1865 till 1868, found it a more dignified, domesticated city, with "long lines of brick stores." He added, "Its dry and wet rivers are newly bridged; irrigating ditches scatter water freely through the streets, lawn and gardens, and now flowers and fruits, trees and vegetables lend their civilizing influences and permanent attractions to the place." Walt Whitman found Denver "well-laid out . . . all the streets with little canals of mountain water running along the sides—plenty of people, `business,' moderness—yet not without a certain racy smack, all its own." Denver was on its way to becoming the "Queen City of the plains and peaks."

Similar development was true in the environs. In 1868 Dr. William Thomas, reporting for the *Rocky Mountain News*, went up the Platte River from Denver toward the mountains. "In riding through this rich and fertile valley, we were struck by the visible signs of improvement on every hand. The log cabins of '59, '60 and '61 stand in marked contrast to the neat and comfortable frame and grout farm houses, within which comfort, contentment and happiness reign. Many farmers are beautifying their places by planting shade and fruit trees."

A crude, raw settlement was on its way to urban respectability and metropolitan status. The population grew from 4,759 in 1870 to 106,713 in 1890, making Denver the nation's fastest-growing city, doubling its population every five years. By 1881 the city had gas, electric light, and telephones, whose poles and wires added "to a metropolitan appearance of the streets." By 1890 Denver was the third-largest city in the West after San Francisco and Omaha, and twenty-fifth in the country. In 1892 Richard Harding Davis, a New Yorker, found it a beautiful city with "the worst streets in the country." They were mud or dust. However, he was struck by the public schools and miles of "the prettiest, strictest, and most proper architecture" of private homes. "They are not merged together on solid rows, but stand apart, with a little green breathing space between, each in its turn asserting it individuality." He also observed the beginnings of a skyline. "Great corporations, insurance

**Photographer Unidentified**
Greeley. *Courtesy of the Denver Public Library Western History Department.*

companies, and capitalists erect twelve-story buildings everywhere. They do it for an advertisement for themselves or their business, and for the rent of the offices." It was not until 1953 that a building over twelve stories was erected.

That skyline now symbolizes Colorado's largest and dominant city. Neal Cassady, the hero of the beat generation, grew up in Denver in the 1930s. His autobiography, *The First Third*, is a superb description of a child's view of the city a half century ago. For a time he lived with his father at the Citizen's Mission on Larimer Street. He recalled his route to Ebert School as a "careful Zig-Zag" of shortcuts and urban exploration:

> I went past some of Larimer's rows of bars and pawn shops, then up 17th Street to the newly-created Federal Reserve Bank with its massive marble squares and with elegant iron bars protecting its windows. Unlike other banks on 17th Street its enormous bronze doors of scrolled bas-relief featuring charioteered archers were never open and I wondered the mystery of its vaults. Another left turn and along a block of Arapahoe Street whorehouses I later patronized. Then a right, onto busy 18th Street with its noisy sheet metal shops and motorcycle showrooms and garages. Across Curtis Street's corner of candy company, parking lot, cheap hotel and cheaper restaurant, and up to Champa Street with the mighty colonnaded structure of the Post Office.... On the Post Office corner I would pause for a quick drink at the public fountain which—unlike most of Denver's—was not shut off in winter so that conical ice attacked the spout's silvery knob, and on certain cold days it victoriously choked the bowl and froze the overflow basin outlet. The idea was to avoid torrents of backlogged water, leap on the fountain to snatch a gulp, then rush a retreat before my shoes were filled. If it had no fresh snow on which to slip, I'd next canter over a huge stone bench whose giant size made even adults use only its edge for sitting. I made paradox of the puzzling proverb carved in its granite, for it cautioned against too much rest while offering it freely, "Desire rest but desire not too much." A springing leap up the 18th Street side of broad stairway that circumvented the Post Office to walk through the warmth of its block-long lobby. On balmier days I disdained this minute of heat, and instead, scampered in a weaving run about the

**Photographer Unidentified**
Cars and Carriages, Walden, ca. 1910. *Courtesy of the Denver Public Library Western History Department.*

fluted sides of every enormous column that fronted Stout Street. There is a hundred feet of sidewalk—filled waste space between curb and Post Office. Down the 19th Street steps three in a bound and catty-corner to a narrow wall's sharp peak which challenged my equilibrium.... Across California Street to go into the alley behind the basement Church of the Holy Ghost, where I later served a year as altar boy without missing a day, then under a vacant lot's billboard to reach the five-pointed intersection of Welton, Broadway and 20th Avenue. The triangular Crest Hotel rose in ten luxurious floors; across from it were a miscellany of drugstore, flower shop, beauty salon, restaurant and two large groceries. Above these businesses there towered, all chunked together, hotel buildings whose solid facades were broken only by the gap of a 20 foot alley entrance. From out of this canyon-like block, so rare in Denver, I raced uphill over another rarity. It was an uncommon pavement bulge with such a stretch from curb to curb that I made a game of traversing its middle.... From the beginning of its swell near Welton Street to its upper end on Sherman, the hill of this 20th Avenue street surface was fully two blocks long and on my path, from Lincoln to Glenarm, was easily a hundred yards in width.... It was as though an entire acre or more had been thoughtlessly razed just so autos could branch out in any one of three directions with unnecessary room to spare, in fact, it had a center triangle, formed by raised oval traffic buttons that contained nothing and was for no purpose but to keep cars out, since some reckless drivers might be tempted into wild and fancy curlyques [sic] over this spacious asphalt—in later years I myself made a few tentative speedway dashes around this triangle. Reaching the sidewalk of Glenarm Street with my lungs in double tempo, I would pass a business school on the corner and then the first of the homes that fringed the downtown area. They nestled between a splendid Catholic church with matching slender spires of stone and the Denver Bible Institute, whose odd belfry was in a squat clapboard affair of afterthought that hid itself beneath the trees of the Institute's yard. Now the residential section began in earnest and from 21st to 22nd Streets only a tiny candy store disturbed the rows of solidly bunched houses. I cut the corner and entered the far side of Ebert's huge graveled playground and ran its length at full tilt for the bell was usually ringing.

Most of Cassady's downtown was razed, to be remade and replaced by shining towers and parking lots. Today's downtown is largely of recent vintage, the product of urban renewal and the energy boom of the seventies, when downtown was known as the "Oil Patch." What remains a half century later? The streets and alleys persist, as well as the bank, the post office, the churches, Ebert School, and the oval traffic lozenges.

## SKI RESORT

At first, skiers were simply brought to the mountains. Accommodations and services followed. For the past thirty years, the Rockies, and Colorado especially, have been the home of new, totally planned resort communities. Although these are predominantly temporary residences, weekend and vacation homes, they are also model communities, having qualities desired not only in a place to visit but also in a more permanent community.

The winter resorts of Vail, Snowmass Village at Aspen, Keystone, Copper Mountain, Beaver Creek, Purgatory, and others were planned as resort "villages." These modern villages are the centers of recreational environments but also represent a nostalgia for a more traditional community, something that never existed in the Colorado Rockies. The prototype is found in the Alps, in a densely packed pedestrian mountain village of traditional architecture, easy access to the slopes, and a communal outdoor space marked by campanile, clock tower, or church steeple. (In the new secular mountain towns, there are no steeples.) This prototype was consciously transported to Colorado's Alpine landscape. Some resort villages were built from scratch, whereas others have been grafted onto the remnants of mining towns, which have found snow, once a nemesis, to be a new source of wealth. Breckenridge is typical. It is two towns. The old Main Street town is now multicolored with postmodern "Victorian" mini-malls—a mining town made charming and quaint. It is coupled with a ski village of high-rise buildings surrounding a lake used for ice skating. The "lake" was a pond left by dredging.

Ski resorts are boomtowns and instant cities. The resort is an intense combination of contemporary, cosmopolitan, urban pleasures in spectacular and wild settings. These are villages of treats, with gourmet food, shopping, entertainment, and the explicit goal of having something for everyone. There are always "things to do," and all paths lead to "The Village." Beyond the center, the pattern is typically rings of condominiums leading to subdivisions of weekend suburbs with their own exurbs.

**GEORGE BEAM**  Cortez with Sleeping Ute Mountain in Background, ca. 1912. *Courtesy of the Colorado Historical Society.*

The path to town and slope is a carefully choreographed procession. Beaver Creek is the newest and most luxurious of the mountain resorts, located off Interstate 70 and linked to Denver, Stapleton Airport, and the rest of the nation. The highway exit is generic exit-world, flanked by a condoland of apartments and shopping centers. An entry gate limits access to the resort valley, a golf course surrounded by luxury homes, many built as modern emulations of the castle-like style of mining "barons" who lorded over communities. The village center is a cluster of great apartment blocks modeled after mountain resort hotels. Ironically, up in the mountains one has a denser, more "urban" experience than in the city. After parking in the underground garage, the visitor takes a passage through the central plaza, whose walkways and staircase, lit by streetlights, lead directly to the slopes. One can ski from mountaintop to town center.

The resort core and the condos are most visible to visitors, but these form only half of the community equation. The urban pattern is reversed. The permanent residents are those who work and service the community. They are of lower economic status and live in new forms of workers' housing. The visitors are the transients. Ski resorts are communities of second homes, which combine the visitor's desire for a wilderness setting with all the modern comforts—taking in a mountain vista while soaking in a Jacuzzi.

The resorts are getaway sites, get-away-from places, as people flee the city and are attracted to these mountain mini-cities. The ski resort has utopian aspirations in its design. There are elements of an "ideal" community in its planned integration of housing, commerce, and recreation. (There is little appearance of "work," and the workers' housing is segregated for economical reasons and to sustain the illusion of a place of pleasure without the conflicts that plague urban life.) All types of community amenities are featured: pools, skating rinks, bike trails, public facilities, and gathering places. Here the idealized time of the "vacation" is given form and substance. Ski resort communities, with their beautifully sculpted ski slopes, have tried to expand their appeal to all four seasons, building golf courses and pools and promoting their facilities as venues for conferences and summer music festivals. Few of these getaway spots have yet evolved beyond the stage of vacation sites, but Vail now has both schools and churches, the early signs of a settled community. There is a fine line between carefully planned villages and places that seem as packaged as the tours

that bring visitors. New problems abound, including haphazard development and increased populations in confined mountain valleys, which are replete with the urban woes of pollution, parking, and traffic jams. Yet the mountains, once home to the makeshift mining camps, now are also subject to the careful orchestration of design guidelines.

**KEN ABBOTT** Marlboro Country, Up-top on the New Colfax Viaduct, March 1984. *Courtesy of the artist.*

**KENNETH HELPHAND**   God's Country Reality, Animas River Valley, 1989. *Courtesy of the author.*

# VISIONS

## "Everyone should exercise scrupulous care."

IN 1869 SAMUEL BOWLES said that in Colorado "lies the pleasure-ground and health home of the nation." More recently Gleaves Whitney has proclaimed the Front Range to be the "Utopia of America." There is a more common vocabulary of idealization, as places in Colorado are often referred to as "God's Country." The Spanish conquerors and explorers, compelled by visions of golden cities, imagined a paradise of material reward. Anglo-American settlers had visions of wealth and prosperity, of finding land and creating community. No one's visions respected the future of the native inhabitants.

Visions are ideals. Some are created; others act as a goal, a chart to guide action. Some visions are illusions, even crackpot schemes of impracticality; others can inspire. Colorado has a history of realized plans and proposals and unrealized dreams, the landscape that might have been. These earlier visions are worthy of reexamination.

**POWELL**

Thomas Jefferson, who died long before "Colorado" existed, perhaps had more to do with its landscape form than any other individual. Jefferson's national ideals are apparent on the land. He purchased Louisiana, absorbing the center of the continent into the nation. He planned the survey system of township, section, and range, the egalitarian and rational grid of squares that structures much of Colorado. In his own writings on Virginia and through his patronage of Lewis and Clark, he set the pattern for nineteenth-century landscape discovery and understanding. It was left to the next generation to realize that the Republic's early desire for a national landscape needed to be tempered by the particular characteristics and needs of its regions. National ideas and ideals necessitated a regional response.

John Wesley Powell, the heroic explorer of the Colorado River, was also the author of a seminal document in American landscape history. *Report on the Lands of the Arid Region of the United States* (1878) was the summary of his experience in the arid West, including Colorado. Powell recognized the distinctive qualities of western lands, where the agricultural and settlement practices of the temperate East needed to be adapted to new geographical circumstances. Wallace Stegner called the report both a program and a manifesto, a prophetic work. West of the hundredth meridian, approximately the line of a predictable twenty inches of annual precipitation, the landscape changed. Powell recommended understanding the limits of this landscape, recognizing western realities, and responding accordingly. Powell's West was not to be an extension of eastern patterns.

Water was the fundamental determinant, but it had to be considered in new proportions. In the East, the 160-acre homestead was an adequate parcel of land, but in the West, 160 acres were too many for an irrigated farm and too few for an unirrigated one. Powell suggested irrigable homesteads of 80 acres and grazing homesteads of 2,560 acres, four full sections. He also urged the grouping of farm residences so "that the inhabitants of these districts [would] have the benefits of the local social organizations of civilization—as schools, churches, etc. and the benefits of cooperation" to counteract the dispersion and scant population. He proposed a revision of the rectilinear land survey system, to change property boundaries to give equal stream rights to all. Fundamental was a land-classification system based on landscape characteristics and not on abstract geometry. Different lands would be treated differently. The system was simple: irrigable lands, found only in relatively small portions in lowlands and around streams; timber lands, mostly found in areas too high or too cold for agriculture; and pasturage lands, found in valleys, mesas, hills, and slopes. Powell proposed careful surveys and a land classification that would identify the best use for the land or soil.

Powell was a strong supporter of irrigation where appropriate and of cooperative labor. He recommended storage reservoirs, dams, and irrigation districts based on watersheds. He proposed irrigation and pasturage districts and advocated water conservation, timber culture, and tragically, a removal of the Indians. Many of his ideas were slowly enacted into law and his methods adopted by the new U.S. Geological and Geographic Survey, an agency

**L. C. MCCLURE**  Barley Raised Without Irrigation, Skelton Ranch, Woodland Park, ca. 1910. *Courtesy of the Denver Public Library Western History Department.*

**SALISBURY**   Camp House, 747 Third Avenue, Durango, 1885. *Courtesy of the Center of Southwest Studies, Fort Lewis College, Durango.*

he would eventually head. Most important, Powell urged a careful reading of the landscape, heeding its lessons and then working within its inherent limits and properties. According to Donald Worster, Powell's vision was "a model of ecological realism in an unsympathetic age of unbounded expectations."

## PARKS

Virtually every Colorado town plan shows land set aside for a park or plaza, a public open space. For example, Denver's first plan had five blocks set aside for public use: a courthouse square, a public square, and three contiguous parcels as a set of park blocks. Why would people laying out a community, confronted with the immediate difficulties of establishing themselves, set aside lands for parks and plazas? Why would they bother, when just beyond the doorstep lay abundant open and virgin lands? These were lands offering much of what we now associate with parks: the wonders of nature, open spaces, and a counterpoint to places of human development.

There are several reasons. Town plans are visions, models of what might be; and by the mid-nineteenth century, the park was considered a basic component of city design. But the park designation also had a symbolic value, a way of placing an "On This Site" sign on the future. It was a designation of expectations, coupled with some trepidation. "Will we make it? Will the town survive?" Of course, many did not. Emblems of the future may have helped as acts of anticipation. "A park is not needed now, but if the town grows, it will be needed for the next generation." Cities were where the movement for public parks began and where parklands were first created. It was only later that the park idea was extended in scale and scope to large nature preserves beyond the urban domain. In Colorado, there was an unusual confluence of events. As one kind of landscape was being changed, others were being set aside to remain unchanged.

Denver, though it came late to park planning nationally, has been a city of innovation and vision. Around 1894 Edward Rollandet, working for the Denver Department of Public Works, planned a park and boulevard system within the structure of the city grid system. In 1906 Charles Mulford Robinson prepared a report to the Denver Art Commission, and the next year this report was augmented by the landscape architect George Kessler for the Denver Park Commission. The Robinson-Kessler Plan adapted well-accepted principles of

park planning to the distinct landscape and city specifics of Denver. The plan used, reinforced, and clarified the city's grid framework, capitalized on vistas of the mountains, exploited the topography, celebrated the preciousness of water in a semiarid land, and linked existing amenities together with new public open spaces. The Robinson-Kessler plan carefully integrated previous park works, notably the superb work of Reinhard Schuetze at Cheesman and Washington parks, and City Park, which Robinson, in Olmstedian vocabulary, called the "People's Park." It was a plan that joined the older open spaces and anticipated future urban expansion through a series of parkways in a "windmill plan" extending east, south, and northwest. In 1929 Saco Rienk DeBoer produced a new master plan for the city, which extended the park and parkway system and projected its expansion far beyond the city limits.

These proposals were all part of a conscious attempt to establish Denver as an exemplar of the City Beautiful movement. A landscape beautiful was a basic tenant of City Beautiful designs. The ideology coupled an expression of great faith in the urban environment as a symbolic presence with the functional efficiency of progressive reformers. There was a strong belief that the physical environment, parks and open spaces especially, exercised a positive moral force. The vocabulary is a bit archaic, but City Beautiful plans combined a sense of civic responsibility with the boosters' optimistic "can-do" spirit.

The most significant fact of the plans was their systemic nature. The result is a park network, an armature that functions as a "structural framework, a design framework, a functional framework, and an aesthetic framework for the city," says Parks Manager Don Etter. The parks and parkways are open and accessible to all, providing places for pleasure and recreation, but they also serve to structure the city, define neighborhood boundaries, and give order, cohesiveness, and identity. They make places for people and celebrate the qualities of the Denver landscape. Golda Meir, who spent two years of her adolescence in Denver, recalled in her autobiography the free concerts in the parks she attended with her husband-to-be. She wrote, "To this day I associate certain pieces of music with the clear, dry mountain air of Denver and the wonderful parks in which Morris and I walked every Sunday in the spring and summer of 1913."

**O. ROACH** Red Rocks, Denver Mountain Parks, ca. 1930. *Courtesy of the Denver Public Library Western History Department.*

**KATHRYN NELSON** The Vega, San Luis, 1980. *Courtesy of the artist.*

Other cities followed suit. In 1912 Charles Mulford Robinson wrote a report for Colorado Springs, and even smaller communities have their elements of the City Beautiful. William Jackson Palmer laid out Durango in 1880 with the Boulevard, now Third Street, as a tree-lined linear park, along with large parks to the north and south of town. These were the foundation blocks for Durango's open-space development.

Another remarkable document was *The Improvement of Boulder, Colorado*, by Frederick Law Olmsted Jr. The report was written in 1910 for the Boulder City Improvement Association, a City Beautiful committee concerned with streets, sidewalks, alleys, sanitation and drainage, tree planting, parks, playgrounds, public education, and "floral culture." Even then, Olmsted noted that people chose to live in Boulder, for it was a "very agreeable locality" where earning money was not primary; it was a place to "lead a satisfactory life." The report dealt with the city streets, waterways, public open spaces, and buildings. Although an outsider from Boston, Olmsted was sensitive to the distinctive aspects of the community: the sharp distinction between the city and the surrounding farmland, wide streets and accompanying alleys, the importance of the sky, and views to the mountains and over the plains. The key component of the proposal was a street layout that gave the city grid a hierarchy and clarity. This was accomplished through detailed attention to the elements of city design: streets and pavements, sidewalks, and street tree planting. Most interesting are Olmsted's comments about grass. He did not want to eliminate lawns but suggested that just bare earth, "raked and swept and kept in order like the floor of a house," was neater and easier to maintain than turf. Olmsted also felt that the irrigation ditches were an "extraordinary opportunity for civic beauty . . . a veritable treasure of municipal decorations, now for the most part neglected and defaced, but all retaining their essential elements unspoiled and ready to shed beauty all about them if only given the proper setting." His recommendations are a primer on the compatibility of utility and beauty.

Denver also had a plan even bolder than its model park and parkway designs. In the first decade of the twentieth century, cars were still a novelty. Denver was a city of only 133,859 in 1900. The idea was proposed for a park system connected to the city but far distant from the city boundaries—the Denver Mountain Parks. Mayor Robert W. Speer in 1907 suggested building an "Appian Way from our city into the mountains." In 1910 a Mountain

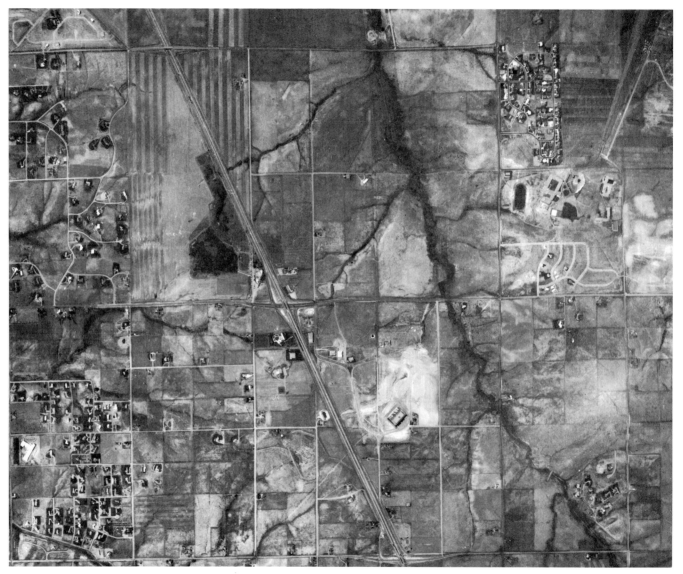

**Photographer Unidentified**  Aerial View of Southeast Denver, August 1964. *Courtesy of Landiscor, Inc., Denver.*

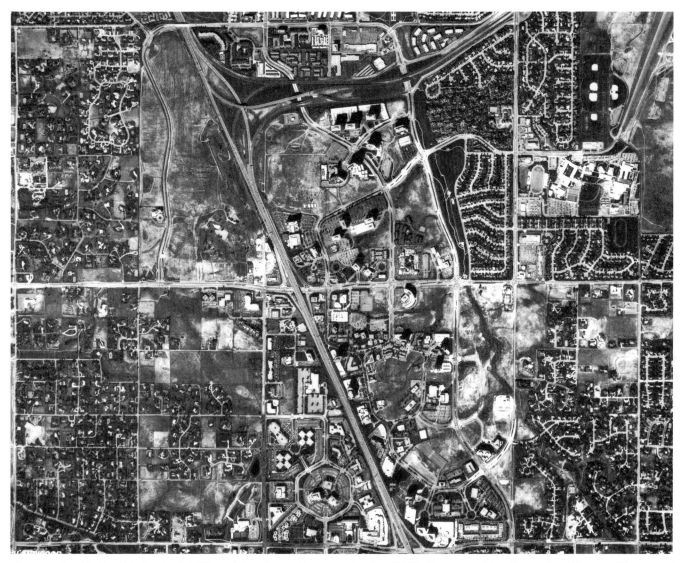

Photographer Unidentified   Aerial View of Southeast Denver, October 1990. *Courtesy of Landiscor, Inc., Denver.* Photograph shows rapid suburban development where small ranches and farms were less than thirty years before. Interstate 25 runs diagonally from top left to bottom right, East Belleview Avenue runs left to right across middle of picture. Denver Technological Center is at bottom center.

**Photographer Unidentified**
Lookout Mountain, Golden and Denver from Wild Cat Point, ca. 1900. *Courtesy of the Denver Public Library Western History Department.*

Parks Committee, with representatives from the chamber of commerce, the real estate exchange, and the motor club, was formed. The Denver Mountain Parks were a regional concept, far beyond the scope of any city in the nation, a system of relatively wild parks, in the foothills of the Rockies, linked by a fifty-five-mile scenic drive. What was remarkable was Denver's early recognition that its hinterlands were integral to the life and structure of the city. The concept also reflected the fact that the automobile had fundamentally changed urban geography and the patterns of leisure. Parks and open spaces were needed not only at one's doorstep, in the neighborhoods, or along the interurban line but also far afield, made individually accessible via roads and cars. The Lariat Trail, which most visitors followed, was well named. Not only did it slither ropelike up Lookout Mountain, but the scenic drives literally roped the foothills and mountains into the Denver area, much as the Denver parkways did at the city scale. The system grew. Winter Park ski area and Red Rocks Park were added. The Red Rocks amphitheater, designed by Burnham Hoyt and the site of summer concerts, is a bold, inspired design that celebrates and enhances one's appreciation of the dramatic hogback landscape. More recently, Denver has recognized the aggregate of street trees, private lawns and gardens, parks and parkways, mountain views, urban vistas, fountains and waterways, and other public open spaces as a "green oasis" in a semiarid landscape and as a "garden system." The garden system concept is a marvelous recognition of the continuity of landscapes from the personal to the public level, from the intimate experience to the distant vista.

As the nation and state became more urban and industrial, the natural world was perceived differently. No longer territory to be settled and domesticated, wildland was deemed worthy of management, protection, and preservation. Barely a generation after the first modern settlers, land was removed from the public domain; in 1891 Congress authorized the president to create forest reserves. In Colorado, the million-acre White River Plateau Reserve was the first in the state, soon followed by the Pike's Peak, Plateau, South Platte, and Battlement Mesa reserves. Theodore Roosevelt added more, and by 1908 over fifteen million acres were part of what was by then a system of National Forests.

Two individuals are deserving of special note for their contributions to landscape conservation. Enos Mills is considered the "Father of Rocky Mountain National Park." Mills

lived at Long's Peak Inn, which he opened in 1902, as innkeeper and mountain guide. More important, he was a publicist, propagandist, and enthusiastic interpreter and protector of the Colorado landscape. He was an uncompromising advocate of National Parks, opposing encroachments on their territory or violations of their spirit. In 1904 he began a campaign for Estes National Park, and eleven years later Rocky Mountain National Park was created.

Arthur Hawthorne Carhart is less known. Carhart was a landscape architect in the Denver office of the U.S. Forest Service. A recreation or landscape engineer, his official job designation, he was informally known as the "beauty engineer." In 1919, after surveying the Trapper's Lake area in the White River National Forest for access and summer home plots, Carhart suggested keeping the area roadless and without built structures. It was, according to Donald Baldwin, the "first de facto application of the wilderness concept," later promulgated by Aldo Leopold and others. Carhart, Baldwin suggests, deserves the title of "Father of the Wilderness Concept." Carhart had recommended a commission to study which areas should be preserved. Trapper's Lake, Colorado's largest natural lake, is now part of the Flat Tops Wilderness Area.

Coloradans are the primary visitors to places that are now on the National Parks cross-country pilgrimage route: Rocky Mountain National Park, Mesa Verde National Park, Great Sand Dunes National Monument, Colorado National Monument, and Dinosaur National Monument. On any day, thousands are hiking in the mountains, rafting rivers, exploring canyons, hunting, fishing, stalking a photograph, identifying wildflowers, or seeking an elusive bird. Visitors arrive singly and in groups of all sizes. The motivations vary: reflection, restoration, education, enjoyment, excitement. From these great natural parks, people return home with memorabilia: photographs, a cherished rock, the memory of a sky never before seen.

## LA VEGA

The "OLDEST TOWN IN COLO" is proudly proclaimed in painted rocks on the hillside above San Luis, overlooking an impressive landscape notable for twin phenomena. First is the 1852 People's Ditch, which still brings water to the valley. Second is La Vega, set aside in 1863. *Vega* is the Spanish word for "meadow," and this flat area must have been a natural meadow, a green place amid an arid landscape, but here *vega* means a commons, set aside in perpetuity. The vega is an *ejido*, common land, which was part of the disputed Sangre de Cristo grant. The 1863 deed gave specific rights regarding pastures, water, wood, and timber. The deed stated, "Everyone should exercise scrupulous care with the use of water without causing harm to their vecinos [neighbors] with the water, nor to anyone; all of the inhabitants shall have with convenient arrangement, the enjoyment of the benefits of the pastures, water, wood and timber, always being careful not to prejudice one another." Members of the plaza were given the right to graze ten cattle or horses on the Vega, which originally encompassed 900 acres. Residents of San Luis have fought to retain their traditional rights, but the courts have whittled away at their claim. Until 1960, the mountain tract was still used as common land, but in 1967 communal grazing rights in the mountains were lost. Of the original acreage, 633.32 acres remain. Unlike the commons of New England communities, La Vega still performs its traditional function as shared grazing land.

What are our landscape ideals? The simple nineteenth-century polarities of civilization and savagery have been replaced by the complex ambiguities of history. Is the frontier still a valid mythology? If the frontier is a spatial edge, a boundary position where cultures meet and clash, what of a myth that emphasizes not the edge but the center? Perhaps the commons can emerge as the new ideal? There may be hope in a return to the first principles. But how do we reconcile frontier and commons? How do we reconcile unlimited possibility, infinite resources, and the escape from responsibility, with the demands of community, with the need to manage, conserve, plan, and share? How can this ideal be reasserted in a diverse multicultural society?

The Vega of San Luis is not a relic, a historical anachronism, but is a touchstone for future possibilities. The commons is shared ground, for a common purpose; ownership is by all, to be commonly and communally managed to protect the resources. The "Tragedy of the Commons," as Garrett Hardin has written, occurs when population pressures on animals, people, or use exceed a certain intensity. Then that commons (the landscape) can no longer carry the load: its capacity is exceeded. Ultimately the commons is destroyed, not in a grand cataclysm but by incremental effects. In Colorado, the tragedy of the commons is the modern dilemma—for the state writ large and for districts and towns. It is certainly so for federal and state lands that are a commons and that need to be understood and managed as such, as well as for water resources, agricultural lands, and historic and cultural resources.

## RE-

re- *prefix* 1. again; anew  2. backward; back

It is time, as Colorado enters the next century, to rethink, revise, redirect, and refocus energies. Where miners claimed, it is now time for reclamation. Where explorers discovered, it is now time for rediscovery. Where people planted, it is now time for replanting. Where there were visions, it is now time for re-vision.

Although pristine mountains and pure streams are still to be found, the frontier ideology, boomtown spirit, and careless and even rapacious practices have soiled, marred, and wasted the Colorado landscape. The challenge is how to imaginatively reclaim these places. From a different viewpoint, landscapes can shift from liability to asset, from neglect to focus, from barrier to connection, from inaccessible to inviting.

For decades, Colorado's urban waterways were allowed to become dumps, engineered conduits, and open sewers. The key connective component of the Robinson-Kessler plan was Cherry Creek Boulevard, later Speer Boulevard. DeBoer proposed exploiting the Platte, Cherry Creek, and gulches as a natural open-space network. It took several generations to build these networks, which are now envisioned as recreational corridors with pathways, bike paths, and boat ramps and, in the most fundamental sense, as waterways once again. In Denver, projects along the Platte River and Cherry Creek are literally redirecting the city to its source, offering amenity, giving structure and order to the city, and capitalizing on this most basic resource. The downtown creekfront of Cherry Creek may take its place as a new "city park" for Denver. The City Beautiful has been abused by automotive excess. The question now is how to reclaim the urban landscape with a new awareness of the city as a natural system. Nature does not reside in the vista or a weekend desire but is immediate. The challenge is to create landscapes that are satisfying habitats, of which we are part. Urban waterways, which capitalize on the dynamic seasonal fluctuations of the semiarid landscape, present the greatest potential. Through mitigation, careful design, and land-use control, they are becoming examples of how a natural system can function as an open-space framework.

The past needs reclaiming as much as landscapes do. When the dominant spirit is one of progress and a desire for the new, it is difficult to retain connections to the past. To preserve the legacy of the state's historic landscape resources, we must recognize these resources as dynamic entities where history is preserved yet where new creation is possible. Colorado's National Historic Districts are selective celebrations and canonizations. Thus far, it is largely the often romanticized history of mining that has been preserved and celebrated, but what of the histories of agriculture, industry, commerce, settlement, or tourism? These landscapes also need attention, care, preservation, and explication. Although preservation activities are to be lauded, the character and the quality of the resulting landscapes need attention. It is difficult to create a "temporal collage," the uneven but exciting and informative layering of historical events, visible in the landscape. In the process of creating new landscapes, too much is selectively stripped away, hidden, destroyed, or buried.

Landscapes can be positively exploited. The daily, the local, and the immediate need equal time with the remote and distant wonders. Access and mobility are still central issues in Colorado, as another layer of linkages is being added to the landscape. There are new trails for walking, hiking, mountain biking, and skiing. These networks, especially in urban areas, are capitalizing on old routes, recognizing the unrealized potential of abandoned railroad beds and canals and recycling these corridors as paths, nature trails, educational opportunities, and components of multiuse open-space systems. The intimacy of the trail reinforces local characteristics and an appreciation of nearby amenities. The High Line Canal, originally begun in 1879, now snakes through suburban Denver, its embankment open for recreational use. At a state level, the 480-mile Colorado Trail from Denver to Durango is slowly being realized.

Changes in conception are critical to landscape understanding and action. Street trees and urban vegetation have been "rediscovered" as part of the urban forest, and groups such as the Denver Urban Forest are replanting lost trees and undertaking efforts at urban reforestation. Portions of the wastelands of "The Great American Desert" are now protected in the Pawnee and Comanche National Grasslands. There has been a gradual realization that all landscapes have their value and need protection and that it is imperative to retain a complete historic landscape legacy for future generations.

What were fundamental truths are recognized as ambiguous assertions. Basic concepts of growth, prosperity, and the "good life" all have a landscape component and are subject to revision. Boom can also be blight; construction is also destruction. The landscape has been a battleground for opposing forces and philosophies. Battles such as those over Two Forks Dam or the 1976 Winter Olympics have crystallized these differing perceptions. Both proposals were defeated. Communities, such as Boulder, have placed limits on growth. The futures of Denver's old and new airports are representative of the great landscape challenges facing the state. The new airport offers the opportunity to design the airport and the dramatic city that will emerge at its borders and to create a gateway that will symbolize the Colorado landscape. The fate of Denver's Stapleton Airport parallels that of the downtown railyards, only now the rate of landscape change has been accelerated. The devolution and the reconstruction of these transportation landscapes present the rare opportunity to reimagine the city and to reconceive growth, transportation, open space, and community design.

Most difficult and dramatic are the emerging visions for reclaiming Rocky Flats and the Rocky Mountain Arsenal. Rocky Flats produced plutonium; the Rocky Mountain Arsenal produced toxic materials, first as a chemical weapons factory, then as a pesticide plant making DDT, chlordane, aldrin, and other products. The arsenal was also where weapons, including mustard gas, phosgene, and bombs, were "unmanufactured" and destroyed. It has the potential to be a model for cleaning up the most insidious of wastes and simultaneously protecting the most sensitive natural resources, for here wildlife undisturbed by direct human contact has taken roost and, remarkably, has survived within the metropolitan habitat of millions of people.

Each generation inherits a landscape. It is a legacy to be squandered or enriched. It is possible to envision a future for the Colorado landscape in which its distinctive qualities are celebrated, its natural cycles are respected, its history is expressed with pride and without ignoring the inherent complexities, and the individual spirit and collective will are fulfilled.

# Bibliography

## General References

Abbott, Carl; Leonard, Stephen J.; and McComb, David. *Colorado: A History of the Centennial State*. Boulder: Colorado Associated University Press, 1976.

Belsse, Robert E., and Goin, Peter, eds. "Nevada's Evolving Landscape." *Nevada Public Affairs Review* 1 (1988).

Boorstin, Daniel. *The Image*. New York: Atheneum, 1976.

———. *The Americans: The Democratic Experience*. New York: Vintage, 1974.

Caughey, Bruce, and Winstanley, Dean. *The Colorado Guide: Landscapes, Cityscapes, Escapes*. Golden, Colo.: Fulcrum, 1989.

Edwards, Mike. "Colorado Dreaming." *National Geographic Magazine* 166, no. 2 (1984).

Erikson, Kenneth A., and Smith, Albert W. *Atlas of Colorado*. Boulder: Colorado Associated University Press, 1985.

Frazier, Ian. *Great Plains*. New York: Penguin Books, 1989.

Garrett, Wilbur E., ed. *Historical Atlas of the United States*. Washington D.C.: National Geographic Society, 1988.

Hart, John Fraser. *Regions of the United States*. Englewood Cliffs, N.J.: Prentice Hall, 1972.

Helphand, Kenneth I. "McUrbia: The 1950's and the Birth of the Contemporary American Landscape." *Places* 5, no. 2 (1988).

Jackson, John Brinckerhoff. *The Southern Landscape Tradition in Texas*. Fort Worth: Amon Carter Museum, 1980.

Kelley, Tim K. *Living in Colorado: A Geography for the Public Schools of Colorado*. Boulder: Johnson Publishing Company, 1951.

Kurtz, Stephen. *Wasteland: Building the American Dream*. New York: Praeger, 1973.

Lee, W. Storrs, ed. *Colorado: A Literary Chronicle*. New York: Funk and Wagnalls, 1970.

Linehan, Edward J. "Colorado: The Rockies Pot of Gold." *National Geographic Magazine* 136, no. 2 (1969).

Lowenthal, David. "The American Scene." *Geographical Review* 58 (1968).

McShine, Kynaston. *The Natural Paradise: Painting in America, 1800-1950*. New York: Museum of Modern Art, 1976.

Marx, Leo. *The Machine in the Garden*. New York: Oxford University Press, 1964.

Meinig, D. W. *Southwest: Three Peoples in Geographical Change, 1600-1970*. New York: Oxford, 1971.

Michener, James. *Centennial*. New York: Random House, 1974.

Mitchell, Robert D., and Groves, Paul A., eds. *North America: Historical Geography of a Changing Continent*. Totowa, N.J.: Rowman and Littlefield, 1987.

Nash, Gerald D. *The American West in the Twentieth Century: A Short History of an Urban Oasis*. Albuquerque: University of New Mexico Press, 1977.

Paul, Rodman W., and Malone, Michael P. "Tradition and Challenge in Western Historiography." *Western Historical Quarterly* 16, no. 1 (1985).

Petulla, Joseph M. *American Environmental History*. San Francisco: Boyd and Fraser, 1977.

Rinehart, Frederick, ed. *Chronicles of Colorado*. Boulder: Roberts Rinehart, 1984.

Smith, Henry Nash. *Virgin Land: The American West As Symbol and Myth*. New York: Vintage, 1950.

Stegner, Wallace. *Angle of Repose*. New York: Doubleday and Company, 1971.

Ubbelohde, Carl; Benson, Maxine; and Smith, Duane A. *A Colorado History*. Boulder: Pruett, 1982.

———, eds. *A Colorado Reader*. Boulder: Pruett, 1982.

Wilkinson, Charles F. *The American West: A Narrative Bibliography and a Study in Regionalism*. Niwot, Colo.: University Press of Colorado, 1989.

*The WPA Guide to 1930's Colorado*. 1941. Reprint. Lawrence: University Press of Kansas, 1987.

Zelinsky, Wilbur. *The Cultural Geography of the United States*. Englewood Cliffs, N.J.: Prentice Hall, 1973.

## Introduction: Landscape Thinking

Clay, Grady. *Close-Up: How to Read the American City*. Chicago: University of Chicago Press, 1980.

Clurman, Harold. "Introducing John Berger." In John Berger, *Toward Reality: Essays in Seeing*. New York: Alfred Knopf, 1962.

Conron, John. *The American Landscape: A Critical Anthology in Prose and Poetry*. New York: Oxford University Press, 1973.

Frye, Northrup. *The Stubborn Structure: Essays on Criticism and Society*. London: Methuen, 1970.

Jackson, John Brinckerhoff. *Discovering the Vernacular Landscape*. New Haven: Yale University Press, 1984.

———. *The Necessity for Ruins, and Other Topics*. Amherst: University of Massachusetts Press, 1980.

———. *Landscapes*. Edited by E. Zube. Amherst: University of Massachusetts Press, 1970.

Lynch, Kevin. *Managing the Sense of a Region*. Cambridge: MIT Press, 1976.

———. *What Time Is This Place?* Cambridge: MIT Press, 1972.

Meinig, D. W. *The Interpretation of Ordinary Landscapes*. New York: Oxford University Press, 1979.

———. "Environmental Appreciation: Localities As a Humane Art." *Western Humanities Review* 25, no. 1 (1971).

Riley, Robert. "Speculations on the New American Landscape." *Landscape* 24, no. 3 (1980).

Shepard, Paul. *Man in the Landscape*. New York: Knopf, 1967.

Stilgoe, John. *The Common Landscape of America, 1500-1845*. New Haven: Yale University Press, 1982.

Tuan, Yi-Fu. *Topophilia*. Englewood Cliffs, N.J.: Prentice Hall, 1974.

Watts, May Theilgard. *Reading the Landscape of America*. New York: Collier Books, 1975.

## Space, Lines, and Sections and Character

Adams, Robert Hickman. *White Churches of the Plains*. Boulder: Colorado Associated University Press, 1970.

Bailey, Robert G. *Description of the Ecoregions of the United States*. Ogden, Utah: Forest Service, U.S. Department of Agriculture, 1978.

Bannon, John Francis. *The Spanish Borderlands Frontier, 1513-1821*. New York: Holt, Rinehart and Winston, 1970.

Benson, Maxine. *Martha Maxwell: Rocky Mountain Naturalist*. Lincoln: University of Nebraska Press, 1986.

Biedahl, A Carl. *New Ground: Western American Narrative and the Literary Canon*. Chapel Hill: University of North Carolina Press, 1989.

Bowles, Samuel. *The Switzerland of America: A Summer Vacation in the Parks and Mountains of Colorado*. Springfield, Mass.: Samuel Bowles and Company, 1869.

Byrd, Gibbens, ed. *This Is a Strange Country: Letters of a Westering Family, 1880-1906*. Albuquerque: University of New Mexico Press, 1988.

Chronic, John, and Chronic, Halka. "Prairie Peak and Plateau: A Guide to the Geology of Colorado" *Colorado Geological Survey Bulletin* 32 (1972).

Eberhart, Perry, and Schmuck, Philip. *The Fourteeners: Colorado's Great Mountains*. Chicago: Swallow Press, 1970.

Eitner, Walter H. *Walt Whitman's Western Jaunt*. Lawrence: Regents Press of Kansas, 1981.

French, Emily. *Emily: The Diary of a Hard-Working Woman*. Edited by Janet Lecompte. Lincoln: University of Nebraska Press, 1987.

Gavin, Jennifer. "Firm's `Human' Maps Plot State's Cultural, Historic Ties." *Denver Post*, June 6, 1989.

Grey, Zane. *Tales of Lonely Trails*. New York: Blue Ribbon Books, 1922.

Griffith, Mel, and Lynell, Rubright. *Colorado: A Geography*. Boulder: Westview Press, 1983.

Hart, E. Richard, ed. *That Awesome Space: Human Interaction with the Intermountain Landscape*. Salt Lake City: Westwater Press, 1981.

Hayden, F. V. *The Great West: Its Attractions and Resources.* Bloomington, Ill.: Charles R. Brodix, 1880.

Ingersoll, Ernest. *The Crest of the Continent.* Chicago: R. R. Donnelley and Sons, 1888.

Jackson, Helen Hunt. *Bits of Travel at Home.* Boston: Roberts Brothers, 1878.

Kendrick, Gregory D., ed. *The River of Sorrows: The History of the Lower Dolores River Valley.* Denver: U.S. Department of the Interior, National Park Service, Rocky Mountain Regional Office, 1982.

Kessler, Edwin, ed. *The Thunderstorm in Human Affairs.* Norman: University of Oklahoma Press, 1973.

Knight, Charles A., and Squires, Patrick, eds. *Hailstorms in the Central High Plains.* Boulder: Colorado Associated University Press, 1982.

La Font, Don. *Rugged Life in the Rockies.* Denver: Big Denver Press, 1966.

Lansford, Henry. "The World above Us: Musings of a Sky-Watcher." *Denver Post Magazine,* Dec. 23, 1984.

Lavender, David. *Colorado River Country.* New York: E. P. Dutton, 1982.

Limerick, Patricia. *The Legacy of Conquest: The Unbroken Past of the American West.* New York: W. W. Norton, 1987.

Luchetti, Cathy. *Women of the West.* St. George, Utah: Antelope Island Press, 1982.

McMechen, Edgar. *The Shining Mountains.* Denver: Denver Public Library, 1935.

May, Stephen. *Pilgrimage: A Journey through Colorado's History and Culture.* Athens: Ohio University Press, 1987.

Milton, John. "Plains Landscapes and Changing Visions." *Great Plains Quarterly* 2, no. 1 (1982).

Momaday, N. Scott, and Muench, David (photographer). *Colorado: Summer/Fall/Winter/Spring.* Chicago: Rand McNally and Company, 1973.

Mutel, Cornelia Fleischer, and Emerick, John. *From Grassland to Glacier: The Natural History of Colorado.* Boulder: Johnson Books, 1984.

Price, Larry W. *Mountains and Man.* Berkeley: University of California Press, 1981.

Rennicke, Jeff. *Colorado Mountain Ranges.* Helena, Mont.: Falcon Press, 1986.

Rockwell, Wilson. *Uncompahgre Country.* Denver: Sage Books, 1965.

Roosevelt, Theodore. *Outdoor Pastimes of an American Hunter.* New York: Scribner, 1982.

Schlissel, Lillian; Ruiz, Vicki; and Monk, Janice. *Western Women: Their Land, Their Lives.* Albuquerque: University of New Mexico Press, 1988.

Simmons, Virginia McConnell. *Valley of the Cranes.* Boulder: Roberts Rinehart, 1988.

Smith, Duane A. "A Land unto Itself: The Western Slope." *Colorado Magazine* 55, nos. 2, 3 (1978).

Sprague, Marshall. *The Great Gates: The Story of the Rocky Mountains Passes.* Lincoln: University of Nebraska Press, 1964.

Stegner, Wallace. *The American West As Living Space.* Ann Arbor: University of Michigan Press, 1987.

Trenton, Patricia, and Hassrick, Peter H. *The Rocky Mountains: A Vision for Artists in the Nineteenth Century.* Norman: University of Oklahoma Press, 1983.

Tweit, Susan. *Pieces of Light: A Year on Colorado's Front Range.* Niwot, Colo.: Roberts Rinehart, 1990.

Webb, Walter Prescott. *The Great Plains.* 1931. Reprint. New York: Grosset and Dunlap, 1971.

Whitney, Gleaves. *Colorado Front Range: A Landscape Divided.* Boulder: Johnson Books, 1983.

Zube, Ervin. "An Exploration of Southwestern Landscape Images." *Landscape Journal* 1, no. 1 (1982).

Zwinger, Ann. *Beyond the Aspen Grove.* Tucson: University of Arizona Press, 1988.

# Exploration

Coves, Eliot, ed. *The Journal of Jacob Flower, 1821-1822.* New York: Francis P. Harper, 1898.

Davidson, Levette J. "Colorado Cartography." *Colorado Magazine* 32, nos. 3,4 (1955).

Frémont, John C. *The Exploring Expedition to the Rocky Mountains*. Washington D.C.: Smithsonian Institution Press, 1964 [1845].

Goetzmann, William. *Exploration and Empire*. New York: Alfred A. Knopf, 1966.

Hafen, Leroy, ed. *Central Route to the Pacific by Gwinn Harris Heap*. Glendale, Calif.: Arthur H. Clark, 1957.

———. *The Journal of Captain John R. Bell: Official Journalist for the Stephen H. Long Expedition to the Rocky Mountains, 1820*. Glendale, Calif.: Arthur H. Clark Company, 1957.

———. *To the Rockies and Oregon, 1839-1842*. Glendale, Calif.: Arthur H. Clark, 1955.

Hafen, LeRoy R., and Hafen, Ann W., eds. *Fremont's Fourth Expedition: A Documentary Account of the Disaster of 1848-49*. Glendale, Calif.: Arthur H. Clark Company, 1960.

———. *The Diaries of William Henry Jackson, Frontier Photographer*. Glendale, Calif.: Arthur H. Clark Company, 1959.

Hayden, F. V. *Geological and Geographical Atlas of Colorado and Portions of Adjacent Territory*. Washington, D.C.: Department of the Interior, 1881.

———. *The Annual Report of the United States Geological and Survey of the Territories Embracing Colorado*. Washington, D.C.: Government Printing Office, 1873.

———. *The First, Second, and Third Annual Reports of the United States Geological Survey of the Territories for the Years 1867, 1868, and 1869 under the Department of the Interior*. Washington, D.C.: Government Printing Office, 1873.

Parkman, Francis. *The Oregon Trail*. Edited by E. Feltskog. Madison: University of Wisconsin Press, 1969.

Pike, Zebulon Montgomery. *The Journals of Zebulon Montgomery Pike*. Edited by Donald Jackson. Norman: University of Oklahoma Press, 1966.

Ruxton, George F. *Adventures in Mexico and the Rocky Mountains*. New York: Harper and Brothers, 1848.

Thayer, William Makepeace. *Marvels of the New West: A Vivid Portrayal of the Stupendous Marvels in the Vast Wonderland West of the Missouri River*. Norwich, Conn.: Henry Bill Publishing Company, 1887.

Thwaites, Reuben Gold ed. *Early Western Travels, 1818-1846*. Vol. 15., *James Account of S. H. Long's Expedition of 1819-1820*. Cleveland: Arthur H. Clark Company, 1905.

Warner, Ted, ed., and Chavez, Fray Angelico, trans. *The Dominguez Escalante Journal: Their Expedition through Colorado, Utah, Arizona, and New Mexicoo in 1776*. Provo: Brigham Young University Press, 1976.

Wilkins, Thurman, with Hinkley, Caroline Lawson. *Clarence King: A Biography*. Rev. ed. Albuquerque: University of New Mexico Press, 1988.

# Rocks

Conner, Daniel Ellis. *A Confederate in the Gold Fields*. Norman: University of Oklahoma Press, 1970.

Dorset, Phyllis Flanders. *The New Eldorado: The Story of Colorado's Gold and Silver Rushes*. New York: Macmillan Company, 1970.

Greever, William. *The Bonanza West: The Story of the Western Mining Rushes, 1848-1900*. Norman: University of Oklahoma Press, 1963.

Hafen, Leroy, ed. *Reports from Colorado: The Wildman Letters and Other Documents*. Glendale, Calif.: Arthur H. Clark Company, 1961.

———. *Colorado Gold Rush: Contemporary Letters and Reports, 1858-59*. Glendale, Calif.: Arthur H. Clark Company, 1941.

———. *Pikes Peak Gold Rush Guidebooks of 1859*. Glendale, Calif.: Arthur H. Clark Company, 1941.

Henderson, Charles. *Mining in Colorado: A History of Discovery, Development, and Production*. U.S. Geological Survey Professional Paper no. 138. Washington, D.C.: Government Printing Office, 1926.

Lewis, Robert S. *Elements of Mining*. New York: John Wiley and Sons, 1964.

McGovern, George, and Guttridge, Leonard F. *The Great Coalfield War*. Boston: Houghton Mifflin, 1972.

Noble, Bruce J. "A National Register Perspective: Evaluating Mining Resources." *CRM Bulletin* 12, no. 2 (1989).

Peterson, Ellen Z. "The Hanging Flume of Dolores Canyon." *Colorado Magazine* 40, no. 2 (1963).

Pritchard, Susan F. *Landscape Changes in Summit County, Colorado, 1859 to Present.* Ph.D. Dissertation, University of Oregon, 1982.

Smith, Duane. *Colorado Mining: A Photographic History.* Albuquerque: University of New Mexico Press, 1977.

## Water

Armstrong, Ellis, ed. *History of Public Works in the United States, 1776-1976.* Chicago: American Public Works Association, 1976.

Davidson, Don. "The Grand River Ditch: A Short History of Pioneering Irrigation in Colorodo's Grand Valley." *Journal of the Western Slope* 1, no. 4 (1986).

Farmer, Edward J. *Colorado's Ground Water Problems.* Water and the Law Bulletin no. 505-S. Fort Collins: Colorado State University Experiment Station, 1960.

Folk-Williams, John A., and Cannon, James S. *Water for the Energy Market: A Sourcebook.* Santa Fe: Western Network, 1983.

Frazier, Deborah. "Colorado: Headwater of the West." *Rocky Mountain News Sunday Magazine,* Oct. 8, 1989.

Friedman, Paul D. "To Tame Cherry Creek: The Birth and Death of the Castlewood Dam." *Colorado Heritage* 1 (1987).

Graf, William L. *The Colorado River: Instability and Basin Management.* Washington, D.C.: Association of American Geographers, 1985.

Hundley, Norris, Jr. *Water and the West: The Colorado River Compact and the Politics of Water in the American West.* Berkeley: University of California Press, 1975.

James, George Wharton. *Reclaiming the Arid West: The Story of the United States Reclamation Service.* New York: Dodd, Mead and Company, 1917.

Knight, Oliver. "Correcting Nature's Error: The Colorado-Big Thompson Project." *Agricultural History* 30, no. 4 (1956).

Lavender, David. *Colorado River Country.* New York: E. P. Dutton, 1982.

MacKendrick, Donald A. "Before the Newlands Act: State Sponsored Reclamation Projects in Colorado, 1888-1903." *Colorado Magazine* 52, no. 1 (1975).

Mellendrie, A. W. "The Hatcher Ditch (1846-1928): The Oldest Colorado Irrigation Ditch Now in Use." *Colorado Magazine* 5, no. 3 (1928).

Radosevich, G. E.; Nobe, K. C.; Allardice, D.; and Kirkwood, C. *Evolution and Administration of Colorado Water Law: 1876-1936.* Fort Collins: Water Resources Publications, 1976.

Reisner, Marc. *Cadillac Desert: The American West and Its Disappearing Water.* New York: Viking, 1986.

Sibley, George. "The Desert Empire." *Harper's* (1977).

Splinter, William E. "Center Pivot Irrigation." *Scientific American* 234, no. 6 (1976).

Trumbo, Dalton. *Johnny Got His Gun.* 1939. Reprint. New York: Bantam Books, 1970.

U.S. Bureau of Reclamation. *The Story of the Colorado-Big Thompson Project.* Washington, D.C.: U.S. Government Printing Office, 1968.

van der Leeden, Frits; Troise, Fred; and Todd, David Keith. *The Water Encyclopedia.* Chelsea, Mich.: Lewis Publishers, 1990.

van der Marck, Jan. "The Valley Curtain." *Art in America,* May-June 1972.

Vranesh, George. *Colorado Water Law.* Boulder: Vranesh Publications, 1987.

Weatherford, Gary D., and Brown, F. Lee, eds. *New Courses for the Colorado River: Major Issues for the Next Century.* Albuquerque: University of New Mexico Press, 1986.

Widtsoe, John A. *The Principles of Irrigation Practice.* New York: Macmillan Company, 1914.

Worster, Donald. *Rivers of Empire: Water, Aridity, and the Growth of the American West.* New York: Pantheon, 1985.

## Plants

American Rock Garden Society and Denver Botanic Gardens. *Rocky Mountain Alpines.* Portland, Oreg.: Timber Press, 1986.

Baker, T. Lindsay. *A Field Guide to American Windmills.* Norman: University of Oklahoma Press, 1985.

Blouet, Brian W., and Lawson, Merlin P., eds. *Images of the Plains: The Role of Human Nature in Settlement.* Lincoln: University of Nebraska Press, 1975.

Davidson, Levette J. "The Festival of Mountain and Plain." *Colorado Magazine* 25, no. 4 (1948).

Dunbar, Robert G. "The Significance of the Colorado Agricultural Frontier." *Agricultural History* 34, no. 3 (1960).

Ebeling, Walter. *The Fruited Plain: The Story of American Agriculture.* Berkeley: University of California Press, 1979.

Fite, Gilbert C. *The Farmers' Frontier, 1865-1900.* New York: Holt, Rinehart and Winston, 1966.

Francis, Mark, and Hester, Randolph T., eds. *The Meaning of Gardens.* Cambridge: MIT Press, 1990.

Fussell, G. E. *Farming Techniques from Prehistoric to Modern Times.* London: Pergamon Press, 1966.

Goff, Richard, and McCaffree, Robert H. *Century in the Saddle.* Denver: Colorado Cattlemen's Association, 1967.

Gras, Norman. *A History of Agriculture in Europe and America.* New York: F. S. Crofts and Company, 1946.

Hart, John Fraser. *The Look of the Land.* Englewood Cliffs, N.J.: Prentice Hall, 1975.

Hewes, Leslie. *The Suitcase Farming Frontier: A Study in the Historical Geography of the Central Great Plains.* Lincoln: University of Nebraska Press, 1973.

Jackson, John Brinckerhoff. *American Space: The Centennial Years, 1865-1876.* New York: W. W. Norton, 1972.

Jefferson, Daisy Baxter. "Pioneer Conditions in the Arkansas Valley." *Colorado Magazine* 24, no. 3 (1947).

Johnson, Hugh. *The International Book of Trees.* London: Mitchell Beasley, 1980.

Klinkenborg, Vern. *Making Hay.* New York: Vintage Books, 1987.

Markoff, Dena S. "A Bittersweet Saga: The Arkansas Valley Beet Sugar Industry, 1900-1979." *Colorado Magazine* 56, nos. 3, 4 (1979).

Peattie, Donald. *A Natural History of Trees of Eastern and Central North America.* Boston: Houghton Mifflin, 1950.

Quillen, Ed. "Lawns: The Easy Solution." *Denver Post,* June 26, 1988.

Ramaley, Francis. *Colorado Plant Life.* Boulder: University of Colorado, 1927.

Rasmussen, Warren D., ed. *Agriculture in the United States: A Documentary History.* New York: Random House, 1975.

Schlebecker, John T. *Whereby We Thrive: A History of American Farming, 1607-1972.* Ames: Iowa State University Press, 1975.

Steinel, Alvin T. *History of Agriculture in Colorado.* Fort Collins: Colorado State Agricultural College, 1926.

Sykes, Hope Williams. *Second Hoeing.* 1935. Reprint. Lincoln: University of Nebraska Press, 1982.

Thayer, Robert. "Beyond Landscape Guilt: Technologies Revolutionize the Landscape." *Landscape Architecture* 74, no. 6 (1984).

U.S. Department of Agriculture. *Fact Book of U.S. Agriculture, 1987.* U.S. Department of Agriculture Miscellaneous Publication no. 1063. Washington, D.C.: U.S. Government Printing Office, 1987.

Wessel, Thomas R., ed. *Agriculture in the Great Plains, 1876-1936.* Washington, D.C.: Agricultural History Society, 1977.

Westermeier, Therese S. "Colorado Festivals." *Colorado Magazine* 30, no. 3 (1953).

## Connections

Beebe, Lucius, and Clegg, Charles. *Narrow Gauge in the Rockies.* Berkeley: Howell-North, 1958.

Belasco, Warren James. *Americans on the Road: From Autocamp to Motel, 1910-1945.* Cambridge: MIT Press, 1981.

Bird, Isabella. *A Lady's Life in the Rocky Mountains.* 1879. Reprint. Norman: University of Oklahoma Press, 1960.

Cafky, Morris. *Colorado Midland.* Denver: Rocky Mountain Railroad Club, 1965.

Christensen, Erin. "Lifelines: The Story of Colorado's Public Road System." *Colorado Heritage,* no. 3 (1987).

Christensen, Erin S., et al. *Challenge to Build: A History of Public Works and APWA in Colorado.* Colorado Chapter, American Public Works Association. Fort Collins: Citizen Print Company, 1987.

*Colorado I-70 Scenic Lands.* Denver: Federal Highway Administration, Bureau of Land Management, and Colorado Division of Highways, 1975.

Cooper, Courtney Ripley. *High Country: The Rockies Yesterday and To-day.* Boston: Little, Brown, and Company, 1926.

Davis, Richard Harding. *The West from a Car-Window.* New York: Harper and Brothers, 1892.

Faithful, Emily. *Three Visits to America.* New York: Fowler and Wells, 1884.

Flink, James. *The Car Culture.* Cambridge: MIT Press, 1975.

Forrest, Kenton, and Albi, Charles. *Denver's Railroads: The Story of Union Station and the Railroads of Denver.* Rev. ed. Boulder: Johnson Publishing Company, 1986.

Gray, D. M., and Male, D. H., eds. *Handbook of Snow: Principles, Processes, Management, and Use.* Toronto: Pergamon Press, 1981.

Greeley, Horace. *An Overland Journal from New York to San Francisco in the Summer of 1859.* 1860. Reprint. Edited by Charles Duncan. New York: Knopf, 1964.

Gregg, Josiah. *Commerce of the Prairies.* 1844. Reprint. Edited by Max L. Moorhead. Norman: University of Oklahoma Press, 1954.

Hafen, LeRoy R. "The Coming of the Automobile and Improved Roads to Colorado." *Colorado Magazine* 7, no. 1 (1931).

Hamil, Harold. *Colorado without Mountains.* Kansas City, Mo.: Lowell Press, 1976.

*Health, Wealth, and Pleasure in Colorado and New Mexico: A Reliable Treatise on the Famous Pleasure and Health Resorts, and the Rich Mining and Agricultural Regions of the Rocky Mountains.* 1881. Reprint. Santa Fe: Museum of New Mexico Press, 1980.

LeMassewa, Robert A. *Colorado Mountain Railroads.* Denver: Sundance Books, 1984.

Liebs, Chester. *Main Street to Miracle Mile.* New York: Little Brown, 1985.

McLuhan, T. C. *Dream Tracks: The Railroad and the American Indian, 1890-1930.* New York: Harry N. Abrams, 1985.

Noel, Thomas. "Paving the Way to Colorado: The Evolution of Auto Tourism in Denver." *Journal of the West* 26, no. 3 (1987).

Ormes, Robert M. *Railroads and the Rockies: A Record of Lines in and near Colorado.* Denver: Sage Books, 1963.

Pikes Peak Ocean to Ocean Highway Association. *General Tourist Guide: Pikes Peak Ocean to Ocean Highway, Los Angeles—New York.* St. Joseph, Mo.: Pikes Peak Ocean to Ocean Highway Association, [c. 1920].

Pomeroy, Earl. *In Search of the Golden West: The Tourist in Western America.* New York: Alfred A Knopf, 1957.

Poor, M. C. *Denver South Park and Pacific.* Denver: Rocky Mountain Railroad Club, 1976.

Rae, John. *The Car Culture.* Cambridge: MIT Press, 1970.

*Rocky Mountain Views on the Rio Grande: Scenic Line of the World.* Denver: Smith Brooks Printing Company, 1914.

Schodek, Daniel L. *Landmarks in American Civil Engineering.* Cambridge: MIT Press, 1987.

Stilgoe, John. *Metropolitan Corridor.* New Haven: Yale University Press, 1983.

"Trails through Time." *Colorado Heritage,* Autumn 1990.

Vaille, Howard T. "Early Years of the Telephone in Colorado." *Colorado Magazine* 5, no. 4 (1928).

Wilkins, Tiris E. *Colorado Railroads.* Boulder: Pruett Publishing Company, 1974.

Williams, Henry T. *The Pacific Tourist.* New York: Henry T. Williams, 1877.

*Wonders of the Rocky Mountains along the Denver and Rio Grande RR.* Denver: Smith Brooks Printing Company, 1920.

## Settlement

Abbott, Carl. "Boom State and Boom City: Stages in Denver's Growth." *Colorado Magazine* 50, no. 3 (1973).

Athearn, Robert G. *The Mythic West in Twentieth-Century America.* Lawrence: University of Kansas Press, 1986.

Banford, Lawrence. "Streets from Silver." *Colorado Heritage,* no. 4 (1987).

Barth, Gunther. *Instant Cities: Urbanization and the Rise of San Francisco and Denver.* New York: Oxford University Press, 1975.

Boyd, David. *A History: Greeley and the Union Colony of Colorado.* Greeley: Greeley Tribune Press, 1890.

Cassady, Neal. *The First Third and Other Writings.* San Francisco: City Lights, 1971.

Colorado State Organization of the International Congress on Tuberculosis. *Colorado Souvenir Book for the International Congress on Tuberculosis.* Denver: Press of Denver Engraving Company, 1908.

Dallas, Sandra. *Colorado Homes.* Norman: University of Oklahoma Press, 1986.

———. *Colorado Ghost Towns and Mining Camps.* Norman: University of Oklahoma Press, 1985.

Dorward, Sherry. *Design for Mountain Communities.* New York: Van Nostrand Reinhold, 1990.

Fairbanks, Robert, and Underwood, Kathleen, eds. *Essays on Sunbelt Cities and Recent Urban America.* College Station: Texas A&M University Press, 1990.

Halaas, David Fridtjof. *Worlds Apart: Indians and Whites in Nineteenth Century Colorado.* Denver: Colorado Historical Society, 1984.

Hamer, David. *New Towns in the New World.* New York: Columbia University Press, 1990.

Hayden, Dolores. *Seven American Utopias: The Architecture of Communitarian Socialism, 1790-1975.* Cambridge: MIT Press, 1976.

Hudson, John C. "Towns of the Western Railroads." *Great Plains Quarterly* (Winter 1982).

Johnson, Clifton. *Highways and Byways of the Rocky Mountains.* New York: Macmillan Company, 1910.

Kerouac, Jack. *Visions of Cody.* New York: McGraw-Hill, 1974.

Lehmann, Susan Collins. *Little Pieces of Time: A Look at Durango's History through Photographs.* Durango: Ore House, 1984.

Malamud, Gary W. *Boomtown Communities.* New York: Van Nostrand Reinhold, 1984.

Miller, Ruth. *Our Trip to Mesa Verde, 1922.* Ouray, Colo.: Buckskin Trading Company, 1988.

Moore, Mechlin D., ed. *Downtown Denver: A Guide to Center City Development.* Washington, D.C.: Urban Land Institute, 1965.

Noel, Thomas J. *The City and the Saloon: Denver, 1858-1916.* Lincoln: University Of Nebraska Press, 1982.

Reps, John. *Cities of the American West: A History of Frontier Urban Planning.* Princeton: Princeton University Press, 1979.

Rohrbough, Malcolm J. *Aspen: The History of a Silver-Mining Town, 1879.* New York: Oxford University Press, 1986.

Scully, Vincent. *Pueblo: Mountain, Village, Dance.* Chicago: University of Chicago Press, 1989.

———. *American Architecture and Urbanism.* New York: Henry Holt and Company, 1988.

Shirvani, H. A., et al. *Urban Design International* 8, no. 1 (1988).

Smith, Duane. *Mesa Verde National Park: Shadows of Centuries.* Lawrence: University Press of Kansas, 1988.

———. "Mining Camps: Myth vs. Reality." *Colorado Magazine* 44, no. 2 (1967).

———. *Rocky Mountain Mining Camps.* Bloomington: Indiana University Press, 1967.

Stoehr, C. Eric. *Bonanza Victorian: Architecture and Society in Colorado Mining Towns.* Albuquerque: University of New Mexico Press, 1975.

Thomas, Lowell. *Good Evening Everybody: From Cripple Creek to Samarkand.* New York: William Morrow, 1976.

Wiley, Peter, and Gottlieb, Robert. *Empires in the Sun: The Rise of the New American West.* New York: G. P. Putnam's Sons, 1982.

## Visions

Abbott, Carl. "The Active Force: Enos A. Mills and the National Park Movement." *Colorado Magazine* 56, nos. 1, 2 (1979).

American Institute of Architecture. *Architecture/Colorado: Mountains, Mines, and Mansions.* Colorado Chapter, American Institute of Architecture, 1966.

Baldwin, Donald N. "Wilderness: Concept and Challenge." *Colorado Magazine* 44, no. 3 (1967).

Carlson. Alvar Ward. "Rural Settlement Patterns in the San Luis Valley: A Comparative Study." *Colorado Magazine* 44, no. 2 (1967).

Chambers, Marlene, ed. *Colorado and the American Renaissance, 1876-1917.* Denver: Denver Art Museum, 1980.

Cutler, Phoebe. *The Public Landscape of the New Deal.* New Haven: Yale University Press, 1985.

DeBoer, S. R. "Plans, Parks, and People." *Green Thumb* 29, no. 5 (1972).
Denver Partnership. *Downtown Creekfront.* Denver: Denver Partnership, 1977.
Etter, Don. "A Legacy of Green: Denver's Park and Parkway System." *Colorado Heritage,* no. 3 (1986).
Etter, Don, and Etter, Carolyn. "Bridging the Turn of the Century: A Vision for Denver." *Green Thumb* 46, no. 1 (1989).
Foster, Mark S. "Denver 76." *Colorado Magazine* 53, no. 2 (1976).
"Is Colorado Helpless against Blight?" *Conservation Foundation Letter,* May 1979.
McHarg, Ian. *Design with Nature.* Garden City, N.Y.: Natural History Press, 1969.
Madson, Chris. *When Nature Heals: The Greening of the Rocky Mountain Arsenal.* Boulder: Roberts Rinehart, 1990.
Martin, Eric. "A Voice for the Wilderness: Arthur H. Carhart." *Landscape Architecture* 76, no. 4 (1986).
Meir, Golda. *My Life.* New York: Putnam, 1975.
Mills, Enos. *Your National Parks.* Boston: Houghton Mifflin, 1917.
———. *Wild Life in the Rockies.* Boston: Houghton Mifflin, 1915.
Moss, Ann. "Along the Scenic Drives and into the Mountain Parks." *Colorado Heritage,* no. 2 (1989).
Nash, Roderick. *Wilderness and the American Mind.* New Haven: Yale University Press, 1973.
Olmsted, Frederick Law, Jr. *The Improvement of Boulder, Colorado.* Boulder: Boulder City Improvement Association, 1910. (Reprinted as Bulletin No. 9, Thorne Ecological Foundation Boulder Company, 1967.)
Powell, John Wesley. *Report on the Lands of the Arid Region of the United States.* 1878. Reprint. Edited by Wallace Stegner. Cambridge: Harvard University Press, 1962.
———. *The Exploration of the Colorado River.* Chicago: University of Chicago Press, 1957.
Robinson, Charles Mulford. *Proposed Plans for the Improvement of the City of Denver.* Denver: Art Commission, City and County of Denver, 1906.
Shoemaker, Joe. *Returning the Platte to the People.* Westminster, Colo.: Tumbleweed Press, 1981.
Singular, Stephen. "Who's King of the Mountain." *Denver Post Magazine,* April 21, 1985.
Smith, Duane A. "Between Boom and Beauty: Durango and the Environment, 1880-1892." *Journal of the West* 26, no. 3 (1987).
Smith, Emilia Gallegos. "Reminiscenses of Early San Luis." *Colorado Magazine* 24, no. 1 (1947).
Smith, Frank E., ed. *Conservation in the United States.* New York: Chelsea House, 1971.
Soergel, Matthew. "Designing Denver." *Rocky Mountain News Sunday Magazine,* March 27, 1988.
Spirn, Anne Whiston. *The Granite Garden: Urban Nature and Human Design.* New York: Basic Books, 1984.
Stegner, Wallace. *Beyond the Hundredth Meridian: John Wesley Powell and the Second Opening of the West.* Boston: Houghton Mifflin, 1954.
Stoller, Marianne, et al. *Las Artistas del Valle de San Luis.* Arvada: Arvada Center for the Arts and Humanities, 1982.
Teeuwen, Randall. *La Cultura Constante de San Luis.* San Luis: San Luis Museum and Commercial Center, 1985.
Weber, David J., ed. *New Spain's Far Northern Frontier: Essays on Spain in the American West, 1540-1821.* Albuquerque: University of New Nexico Press, 1979.
Wilson, William A. "New Wine in Old Bottles: The Denver Mountain Parks Movement." *Colorado Heritage,* no. 2 (1989).

# Photography

*The references cited here are intended to provide a broad context for the history of landscape photography in Colorado. Although several general texts on photography in the American West have been included, the selection is focused primarily on work that addresses the human interaction with the Rocky Mountain/Great Basin landscape. The reader is encouraged to further consult the hundreds of smaller monographs, catalogues, and regional publications that have been produced by individuals and local and regional historical societies.* —E. M.

Adams, Ansel, and Alinder, Mary Street. *Ansel Adams: An Autobiography.* Boston: Little, Brown, and Company, 1985. See also the numerous monographs and books by and about Ansel Adams.

Adams, Robert. *To Make It Home: Photographs of the American West.* New York: Aperture, 1989.

———. *From the Missouri West: Photographs by Robert Adams.* Millerton, N.Y.: Aperture, 1980.

———. *Prairie: Photographs by Robert Adams.* Denver: Denver Art Museum, 1978.

———. *The New West: Landscapes along the Colorado Front Range.* Boulder: Colorado University Press, 1975.

Alinder, Mary Street, and Stillman, Andrea Gray, eds. *Ansel Adams: Letters and Images, 1916-1984.* Boston: Little, Brown, and Company, 1988.

Bartlett, Richard A. *Great Surveys of the American West.* Norman: University of Oklahoma Press, 1962.

Bright, Deborah. "Of Mother Nature and Marlboro Men: An Inquiry into the Cultural Meanings of Landscape Photography." *Exposure* 23, no. 4 (1985).

Carvalho, Solomon N. *Incidents of Travel and Adventure in the Far West: With Col. Fremont's Last Expedition across the Rocky Mountains, Including Three Months' Residence in Utah, and a Perilous Trip across the Great American Desert to the Pacific, by S. N. Carvalho, Artist to the Expedition.* New York: Derby and Jackson, 1857.

Combs, Barry B. *Westward to Promontory: Building the Union Pacific across the Plains and Mountains.* Photographs by Andrew J. Russell. New York: Garland Books, 1969.

Current, Karen, and Current, William R. *Photography and the Old West.* New York: Harry N. Abrams, 1978.

Dawson, Robert. *Robert Dawson Photographs.* Introduction by Ellen Manchester. Tokyo: Min Gallery, 1988.

Finkhouse, Joseph, and Crawford, Mark, eds. *A River Too Far: The Past and Future of the Arid West.* Photographs by the Water in the West Project. Reno: Nevada Humanities Committee and the University of Nevada Press, 1991.

Hafen, LeRoy R., and Hafen, Ann W., eds. *The Diaries of William Henry Jackson, Frontier Photographer.* Glendale, Calif.: Arthur H. Clark Company, 1959.

Hagen, Charles, ed. "Beyond Wilderness." *Aperture* 128 (Summer 1990).

Harper, Opal M. "A Few Early Photographers of Colorado." *Colorado Magazine* 33, no. 4 (1956).

Hayden, F. V. *Sun Pictures of Rocky Mountain Scenery, with a Description of the Geographical and Geological Features, and Some Account of the Resources of the Great West; Containing Thirty Photographic Views along the line of the Pacific Rail Road, from Omaha to Sacramento.* Photographs by A. J. Russell. New York: Julius Bien, 1870.

———. *Annual Reports of the United States Geological Surveys of the Territories.* Washington, D.C.: Government Printing Office, 1868-83.

Holborn, Mark, ed. "Western Spaces." *Aperture* 98 (Spring 1985).

Horan, James D. *Timothy O'Sullivan: America's Forgotten Photographer.* Garden City, N.Y.: Doubleday, 1966.

Hume, Sandy; Manchester, Ellen; and Metz, Gary. *The Great West: Real/Ideal.* Introduction by Nathan Lyons. Boulder: University of Colorado, 1977.

Jackson, Clarence S. *Picture Maker of the Old West, William H. Jackson.* New York: Charles Scribner's Sons, 1947.

Jackson, William Henry. *Time Exposure: The Autobiography of William Henry Jackson.* New York: G. P. Putnam's Sons, 1940.

———. *The Pioneer Photographer: Rocky Mountain Adventures with a Camera.* Yonkers-on-Hudson, N.Y.: World Book Company, 1929.

———. *Descriptive Catalogue of the Photographs of the United States Geological Survey of the Territories, for the Year 1869 to 1875, Inclusive.* Miscellaneous Publications No. 5. Washington, D.C.: Government Printing Office, 1875.

Jenkins, William. *New Topographics: Photographs of a Man-Altered Landscape.* Rochester, N.Y.: International Museum of Photography at George Eastman House, 1975.

Johnson, Denis, and Galassi, Peter. *Traces of Eden: Travels in the Desert Southwest.* Photographs by Mark Klett. Boston: David R. Godine, Publisher, 1986.

Jones, William C., and Jones, Elisabeth. *Photo by McClure: The Railroad, Cityscape, and Landscape Photographs of L. C. McClure.* Boulder: Pruett Publishing Company, 1983.

Jussim, Estelle, and Lindquist-Cock, Elizabeth. *Landscape As Photograph.* New Haven: Yale University Press, 1985.

Keller, Ulrich. *The Highway As Habitat: A Roy Stryker Documentation, 1943-1955.* Santa Barbara, Calif.: University Art Museum, 1986.

King, Clarence. *Annual Reports of the United States Geological Explorations of the Fortieth Parallel.* Washington, D.C.: Government Printing Office, 1871-78.

Klett, Mark; Manchester, Ellen; and Verburg, JoAnn. *Second View: The Rephotographic Survey Project.* Albuquerque: University of New Mexico Press, 1984.

Lindquist-Cock, Elizabeth. *The Influence of Photography on American Landscape Painting, 1839-1880.* New York: Garland, 1977.

Mangan, Terry William. *Colorado on Glass: Colorado's First Half Century As Seen by the Camera.* Denver: Sundance Limited, 1975.

Naef, Weston. *Era of Exploration: The Rise of Photography in the American West, 1860-1885.* Boston: New York Graphic Society, 1975.

Newhall, Beaumont. *The History of Photography from 1839 to the Present Day.* New York: Museum of Modern Art, 1949.

Newhall, Beaumont, and Edkins, Diana. *William H. Jackson.* Dobbs Ferry, N.Y.: Morgan and Morgan, 1974.

Ostroff, Eugene. *Western Views and Eastern Visions.* Washington, D.C.: Smithsonian Institution, 1981.

Ott, Richard, ed. *When the River Was Grand: Historical Views of Western Colorado's Grand Valley.* Grand Junction: Museum of Western Colorado, 1982.

Paddock, Eric. "One Man's West: The Photography of Timothy H. O'Sullivan." *Colorado Heritage*, no. 2 (1986).

Plowden, David. *Commonplace.* New York: Sunrise, 1974.

———. *The Hand of Man upon America.* New York: Braziller, 1974.

Russell, Andrew J. *The Great West Illustrated in a Series of Photographic Views across the Continent, Taken along the Line of the Union Pacific Railroad, West from Omaha, Nebraska, vol. 1.* New York: Union Pacific Railroad, 1869.

Snyder, Joel. *American Frontiers: The Photographs of Timothy O'Sullivan, 1867-1874.* Millerton, N.Y.: Aperture, 1981.

Taft, Robert. *Photography and the American Scene: A Social History, 1839-1889.* New York: Macmillan Company, 1938. Reprint, New York: Dover Publications, 1964.

Thayer, William Makepeace. *Marvels of the New West: A Vivid Portrayal of the Stupendous Marvels in the Vast Wonderland West of the Missouri River.* Norwich, Conn.: Henry Bill Publishing Company, 1887.

Wheeler, George M. *Report upon United States Geographical Surveys West of the One Hundredth Meridian.* 7 vols. Washington, D.C.: Government Printing Office, 1875-89.

Wolf, Daniel. *The American Space: Meaning in Nineteenth-Century Landscape Photography.* Introduction by Robert Adams. Middletown, Conn.: Wesleyan University Press, 1983.

Yates, Stephen A., ed. *The Essential Landscape: The New Mexico Photographic Survey.* Introduction by J. B. Jackson. Albuquerque: University of New Mexico Press, 1985.

# Index

Adams, Robert xvii, 57
Agriculture 19,122, 143-159; dryland farming 151; experiment stations 151; suitcase farming 156; technology 141-151
Air Force Academy 211
Akron 151
Alamosa 43, 235,
Amache 228
Anasazi 119, 121, 219, 231
Andrews, Darwin 174
Anthony, Webster 91
Antonito 143, 203
Arapahos 11, 55 110, 218
Arch, Matt 125
Argo 103
Arkansas River 5, 9, 13, 15, 69, 73, 75, 89, 133, 137, 218
Arkansas River Valley 15, 19, 35, 39, 75, 113, 147, 183, 201,
Aspen 57, 171, 193, 229
Aspens 57, 169, 171, 193
Athearn, Robert 55, 236
Auraria 93, 239
Aurora 229
Austin 151

Baca County 156
Baldwin, Donald 262
Barbed wire 165
Barth, Gunther 237
Battlement Mesa 229
Beale Expedition 79
Beaver Creek 192, 243, 245
Bell, James 71
Bell, John 15, 29, 69, 71

Benton, Thomas Hart 38, 75
Bent's Fort 89, 121, 169
Big Thompson Project 135-137
Big Timbers 89, 169
Bijeau, Joseph 39
Bird, Isabella 37, 57, 197, 207, 211
Bison 11, 73, 143, 159
Black Canyon of the Gunnison 25, 129
Black Hawk 85, 179, 206
Black Sunday 109
Bliss, Edward 121, 145, 162
Bonanza 231
Boom and bust 49-53, 229-231
Borderlands 19
Boulder 43, 79, 169, 174, 237, 257
Bowles, Samuel 31, 35, 38, 39, 57, 113, 147, 239, 249
Breckenridge 53, 85, 99, 192, 213, 229, 243
Brewer, William 231
Brown, Maggie 231
Buckskin Joe 85, 231
Buena Vista 159, 183, 189, 206
Bureau of Land Management 167, 203
Bureau of Reclamation 127, 135, 139, 140; Uncompaghre Irrigation Project 129
Burke, Edmund 59
Burlington 189, 228
Byers, William 87

California Gulch 105
Cameo 26
Cameron, Robert A. 123, 235
Camp Hale 213, 228
Camps 221-228; autocamp 229; Health camps 228

Canals 23, 121-123, 157; Acequia Madre 121; Denver High Line Canal 123, 265; Grand Valley Canal 125; Larimer and Weld Canal 123; North Poudre, Bessemer, Fort Lyon, Bob Creek, and Otero 123
Cañon City 73, 109, 179, 189, 201, 203, 206
Caracas 68
Carbondale 43, 229
Carhart, Arthur Hawthorne 262
Carson, Kit 75
Carvalho, Solomon N. xv-xvi
Cascade 206
Cassady, Neal 241
Centennial 133, 236
Central City 85, 95, 99, 185, 206
Chama 121
Cherry Creek 38, 85, 93, 137, 264
Cheyenne Wells 71, 151
Cheyennes 11, 55, 110, 151
Chinese 109
Cíbola 85
City Beautiful 255, 257
Climate 5, 19, 29, 31, 117, 121, 153, 215
Climax Mine 64
Cokedale 101, 109
Collier, D.C. 93
Colonies 232-235; Chicago-Colorado Colony 123, 235; Fort Amity 235; Fort Collins Agricultural Society 123; Fountain Colony 235; St. Louis Western Colony 123; Tennessee Colony 123; Union Colony 109, 123, 232-235
Colony Project 109, 229
Color 11, 55-59, 85
Colorado City 95

Colorado National Monument 262
Colorado Plateau 5, 68, 95, 113
Colorado River 113, 129, 135, 251; Colorado River Compact 135, 139; Grand River 123
Colorado School of Mines 43
Colorado Springs 12, 21, 43, 189, 192, 201, 206, 209, 235, 257
Colorado Territory 69
Colorado Trail 265
Cooper, Courtney 59, 189
Cortez 43
Cotopaxi 235
Cottonwoods 7, 13, 169, 225
Craig 26, 229
Creede 87, 95, 107, 179, 223
Crested Butte 179, 213
Cripple Creek 85, 95, 183, 206
Crystal 211, 231
Cuchara 201
Curtis, Samuel 38

Dams 64, 129, 133, 137, 139-140; Blue Mesa 137; Castlewood 137; Cheesman 131; Lawn Lake 137; Narrows 140; Two Forks 140, 266
Davis, Richard Harding 97, 201, 223, 239
DeBoer, Saco Rienk 255, 264
Deere, John 147
Deerfield 235
Del Norte 43
Delta 26
Denver 31, 35, 48, 53, 64, 85, 89, 93, 122, 145, 165, 169, 174, 183, 185, 192, 197, 201, 211, 225, 237-243, 245, 253-260, 264; airport 192, 266; Denver Mountain Parks 45, 189, 203, 237, 257; parks 37, 174, 229, 253; Robinson-Kessler plan 253, 264; transit systems 183; Union Station 45; water supply 131-137, 140
Desert Land Act 153
Dillon 133, 179
Dinosaur National Monument 140, 262
Ditches 9, 95, 127, 151; Cameron Pass Ditch 131; Consolidated Ditch 99; Grand River Ditch 124; Hatcher Ditch 121; Pacific Slope Ditch 123; People's Ditch 262; Pioneer Ditch 123; San Luis People's Ditch 121; San Pedro Ditch 121; Skyline Ditch 131
Dolores Canyon 99
Domínguez-Escalante Expedition 68
Durango 109, 141, 183, 193, 203, 206, 235, 257
Dust Bowl 156

Edgewater 228
El Moro 235
Eldora 179
Elevation 5, 31, 113
Ephraim 235
Estes Park 35, 135, 141, 197, 203, 206
Etter, Don 255

Fairplay 85, 104
Faithful, Emily 181, 183
Feedlot 15, 167
Felker, W.B. 187
Ferril, Thomas Hornsby 115
Festivals 159; Lamb Day 159; Lettuce Day 159; Melon Day 159; Peach Day 159; Pickle Day 159; Potato Days 159; Sand and Sage 15; Strawberry Day 159; Tomato Day 159
Firstview 71
Fort Collins 123, 151, 156, 159, 174, 183, 192, 236
Fort Lewis 151
Fort Lupton 159
Fountain 235
Frémont, John C. 29, 31, 33, 39, 75, 78, 169
Front Range 5-9, 19-23, 31, 33, 39, 57, 73, 115, 140, 229, 241, 244
Frontier 49, 53, 264
Fruita 26, 125, 159
Fruitvale 159

Garden of the Gods 57, 197, 211
Gardens 171-174, 260
Garrard, Lewis 143
Garrison, A.F. 91
Gateway 117
Geology 11, 23, 85, 95
Georgetown 85, 95, 181, 199
Ghost towns 53, 229-233
Glenwood Canyon 192
Glenwood Springs 159, 189, 206
Glidden, J. F. 165
Golden 145, 187, 203
Granada 228
Grand Junction 26, 109, 123, 125, 159, 189, 237
Grand Valley 123, 127, 159, 183
Granite 79
Grant, Ulysses 122
Graymont 179
Great American Desert 71, 145, 147, 162, 173, 265
Great Sand Dunes National Monument 262
Greeley 13, 122, 123, 159, 179, 183, 187, 233
Greeley, Horace 37, 91, 122, 197, 239
Gregory diggings 87, 91, 105, 197, 221
Gregory, John 91, 145
Grey, Zane 57, 231
Grover 31
Gulch 97
Gunnison 43, 109, 179, 183, 206

Hamer, David 233
Hamil, Harold 13, 151, 185, 193
Hardin, Garrett 263
Harrison, Charles 131
Hatch Act 151
Hatcher's farm 143
Hawken, Sam 113
Hayden, Ferdinand V. 35, 57, 79, 95
Hayden Survey 80, 122
Heap, Gwinn Harris 39, 79
Held, Phillip 153
Henderson, Charles 109
Hill, Nathaniel 103
Hole-in-the-Prairie 121
Hollister, O. J. 105

# Index

Holly 165
Homestead Act 153, 251
Hooper, Shadrach K. 199
Hotels 35, 206
Hoyt, Burnham 260
Hubbard, Elbert 159
Hydraulic civilization 115, 117

Idaho Springs 85
Instant cities 221, 237
Ironton 179
Irrigation 15, 115-130, 157, 235, 251, 257; center-pivot irrigation 130; statistics 125, 130

Jackson, Helen Hunt 37, 199, 221
Jackson, O. T. 235
Jackson, William Henry xvi, 80, 203, 219
James, Edwin 69, 71, 73, 113
James, George Wharton 129
Jefferson, Thomas 64, 249

Kessler, George 253
Keystone 192, 243
Kidd, W.H. 173
Kiowa County 156
Klinkenborg, Vern 147
Knight, Oliver 137

La Junta 19
La Veta 235
Lake City 206
Lamar 15
Landscape xxi-xxvi
Larimer, William H. 93
Lawns 171, 257
Leadville 85, 89, 95, 103, 179, 189, 206, 221
Leopold, Aldo 262
Limon 189
Logan County 13
Long Expedition 15, 39, 69, 113, 225
Long, Stephen 71
Longmont 43, 123, 179, 235

Loveland 156
Ludlow 110
Lynch, Kevin 55

Main Street 193, 215, 223, 235-237
Manassa 5, 235
Manitou 197, 201, 206, 211
Manitou Springs 79, 209
Maps 80
Maroon Bells 211
May, Stephen 33
Maybell 109
McClure 171
McCormick, Cyrus 147
Mead, Elwood 125
Meeker 109, 123, 229, 236
Meeker, Nathan 122, 125, 232
Meers, Otto 185
Meir, Golda 255
Mesa Verde 121, 179, 211, 218, 262
Michener, James 133, 236
Mills, Enos 260
Mining 85-110, 199, 265; claims 97; diggings 95, 105; hydraulic 99; placer mining 97; railroads 103; rushes 85-95, 223; sounds 103; subsurface mining 103; techniques 99-103
Mining camps 53, 91-95, 221-225, 229-231
Momaday, N. Scott 33, 37
Monotony 71
Monte Vista 151
Montezuma 229
Montrose 26, 129, 183, 189
Monument 235
Mormons 26, 129, 183, 189
Morrison 203
Morrow Point 26
Mountain men 78
Mountains 5, 7, 19-25, 33-43, 59, 69, 75, 85, 101, 211, 213; Aspen Mountain 43; Culebra Range 25; Fisher Peak 43; Flatirons 43; Flattop 57; "Fourteeners" 38; Gray's Peak 38; Holy Cross 211; James Peak 183; Long's Peak 33, 43,

197; Lookout Mountain 189, 203, 260; Medicine Bow 23; Mount Blanca 43; Mount Evans 179; Mount McClennan 201; Mount Princeton 211; Mount Sopris 43; Mount Wilson 43; Park Range 23; Rabbit Ears 23; San Juan 68, 87, 103, 185, 203; Sangre de Cristo 25, 33; Silver Mountain 33; Sleeping Ute 43; Spanish Peaks 25, 39, 43; Wet 39

Native-Americans 19, 119, 159; settlements 218
Naturita 109
Newlands Act 129

Oakley, Obadia 39, 71
Ogallala aquifer 15, 130
Oil shale 87, 109
Olmsted Jr. Frederick Law, 257
Orchard 123, 236
Oregon Trail 15, 89, 169, 177
Oro City 85, 91
Osgood, John 235
Otero County 159
Ouray 87, 103, 179, 185, 203, 206

Palisade 125
Palmer (Arkansas) divide 15, 57
Palmer, William Jackson 257
Paonia 26
Parachute 229
Parkman, Francis 33, 37, 177, 218
Parks 22, 39, 59, 78; Bergen Park 23; Egeria Park 23; Estes Park 23; Medicine Bow 23; Middle Park 23, 59; North Park 23; Poudre Park 23; San Luis Valley 25, 115; South Park 39, 85, 179, 181, 231; Wet Mountain Valley 23; Woodland Park 23
Parsons, William 29, 39
Passes 43; Altman 183; Cuchara 206; Cumbres 203; Kenosha 181, 185; Loveland 192; Mosquito 89; Raton 162; Vail Pass 192
Photography xv-xviii, 75, 80, 206
Piedmont 109, 122, 232

Pike, Zebulon 38, 59, 80
Pike's Peak 21, 38, 43, 71, 73, 79, 179, 187, 189, 203, 211, 260
Plains 3, 11-15, 71, 171, 236
Platte River 5, 9, 13, 71, 78, 93, 137, 218, 239, 264
Platte River Valley 13, 15, 19, 78, 119, 147
Platteville 159
Plazas 219
Poncha Springs 229
Powell, John Wesley 249-253
Prairie 11, 147, 159
Preuss, Charles 75
Proctor 195
Prowers County 15
Prowers, John W. 165
Pueblo 21, 141, 174, 179, 201, 237

Quillen, Ed 171
Quivira 85

Railroads 13, 38, 55, 79, 103, 179-185, 206, 207; Argentine Central 201; Atchison, Topeka and Santa Fe 201, 215; Burlington Northern 13; Cumbres and Toltec 203; Denver and Rio Grande 129, 185, 199, 203, 207, 215, 223; Denver Pacific 179; excursion trains 195, 201; Georgetown Loop 181, 201; interurbans 183; Kansas Pacific 71, 162, 179; Manitou and Pike's Peak Cog Railway 181; Narrow-gage 181; Santa Fe 162; Union Pacific 13, 185
Ranching 159-167; cattle trails 162; open range 162; sheep 165; statistics 165-167; stock law 163
Rangely 87, 229
Red Cliff 57
Red Rocks 57, 260
Redmesa 57
Redstone 57, 235
Reisner, Marc 115, 139

Republican River Trail 89
Reservoirs 127; Arapahoe 135; Carter Lake 135; Dillon 133, 135; Green Mountain 135; Horsetooth Reservoir 135; John Martin 133; Lake Granby 135; McPhee 139; Shadow Mountain 135; Willow Lake 135
Richardson, Albert D. 91
Richfield 235
Rifle 109
Rifle Gap 135
Rivers: Animas 113; Big Sandy Creek 89; Blue 101, 113, 133, 135; Boulder Creek 135; Cache la Poudre 122, 135, 235; Clear Creek 145; Costilla 121; Culebra 121; Dolores 113, 131; Eagle 131; Elk 23; Fraser 133; French Creek 101; Green 113; Greenhorn Creek 78; Gunnison 25, 113, 129, 133; Horsefly Creek 68; Huerfano 39, 113; Laramie 131; Missouri 113; Montezuma 131; Monument Creek 79, 81; Purgatory 78, 121, 143; Rio Grande 5, 113, 179; Saguache Creek 113; San Miguel 99; South Boulder Creek 133; St. Vrain Creek 135; Swan 101; Uncompaghre 131; West Plum Creek 57; Yampa 23, 113
Roads 25, 185-195; alleys 193; automobile trails 189; highway strip 206; I-70 192; Interstate Highway System 189, 192; Million Dollar Highway 185, 203; Pike's Peak Ocean-to-Ocean Highway 189; scenic drives 203, 260; Skyline Drive 35, 211; Trail Ridge Road 35, 203; U.S. Highway 50 71
Robinson, Charles Mulford 253, 257
Rock Creek 179
Rocky Flats 64, 266
Rocky Ford 151, 159
Rocky Mountain Arsenal 266
Rocky Mountain National Park 35, 135, 189, 203, 262
Rocky Mountain News 87, 91, 121, 239
Rocky Mountains 5-9, 23, 33-45, 69-80, 85, 87, 113, 179, 197, 243

Rollandet, Edward 253
Roosevelt, Theodore 25, 260
Royal Gorge 73, 179, 181, 183, 203, 211
Russell Gulch 99
Russell, William Green 85, 99
Ruxton, George 33, 78, 121, 169, 225

Saints John 229
Salida 43, 189, 231
San Luis 122, 143, 262; La Vega 262; San Luis People's Ditch 121
Sand Creek Massacre 110, 228
Sanford 235
Sangre de Cristo grant 219, 263
Santa Fe Trail 15, 89, 169
Scenery 3, 33-38, 68, 73, 199-201
Schlebecker, John T. 156
Schuetze, Reinhard 255
Seibert 228
Seymour, Samuel 71
Silver Plume 87, 181
Silverton 43, 87, 103, 179, 185, 203, 206, 236
Simla 228
Skiing 207, 211-213, 243-247
Sky 11, 31, 257
Skyline 11, 33, 239
Slick Rock 109
Smith, Duane 201
Smoky Hill Trail 89
Somerset 179, 229
Spanish 19, 68, 85, 119, 163, 249
Sparks, Felix 117
Speer, Robert W. 257
Sprague, Marshall 43, 80, 185
St. Charles 93
Steamboat Springs 207
Stegner, Wallace 57, 215, 236, 251
Steinel, Alvin 156
Sterling 174
Strasburg 130
Stratton 238

# Index

Sugar beets 15, 156-159
Summit County 99
Swink, George 159
Sykes, Hope Williams 157, 236

Table Rock 151
Taft, William Howard 129
Tailings 105, 110, 192
Tarryall 85
Taylor, Bayard 221
Taylor, Edward T. 167
Taylor Grazing Act 167
Telegraph 89
Telephone 89, 239
Telluride 53, 87, 103, 193, 203, 213
Territory of Jefferson 69
Thomas, Cyrus 122
Thomas, William 239
Tierney, Luke 207
Timber and Stone Act 153
Timber Culture Act 153
Timberline 7, 9, 21, 31
Toltec Gorge 203, 211
Tourism 55, 183, 195-211, 243
Towns 9, 11, 21, 45, 55, 91, 103, 171, 215
Trapper's Lake 262
Trinidad 43, 101, 179, 183, 193, 201

Trumbo, Dalton 125
Tunnels: Alpine 183; Alva B. Adams 135; Eisenhower 192; Gunnison 129; Harold D. Roberts 133; Moffat 133, 183
Turner, Frederick Jackson 45
Tweit, Susan 57

U.S. Forest Service 167, 203, 260
Uncompaghre Valley 129
Uranium 87, 109, 231
Uravan 109, 231
Urban parks 253-260, 265
Ute 55, 68, 104, 133

Vail 192, 213, 243
Vail, Charles D. 192
Valley Curtain 133
Valmont 79
Vernal Mesa 129
Villard, Henry 91

Wagon Wheel Gap 179, 223
Walden 237
Wall, David K. 145
Walsenburg 43, 235
Water 113-141, 251; diversion 131-137; floods 137; groundwater 130; statistics 115, 125, 130

Water law 137-140; Colorado Doctrine 138
Webb, Walter Prescott 115
Weld County 15, 123, 127, 167
Weldona 140
Wellington, John 125
Westcliffe 179
Western Colorado State College 43
Western Slope 11, 25, 53, 109, 131, 159, 236
Wetherill Mesa 121
Whitman, Walt 57, 239
Whitney, Gleaves 249
Wilderness 49, 177, 262; Flat Tops Wilderness Area 262
Wildlife 11, 68, 73, 201
Williams, Henry 199
Windmills 11, 130, 149
Winter Park 207, 260
Witfogel, Karl 115
Worster, Donald 115, 117, 138, 253

Xeriscape 141, 174

Yampa 231
Yont, C. A. 187

Zybach, Frank 130

# Index to Photographers

Abbott, Ken 246
Adams, Ansel 4
Adams, Robert 8, 12, 30
Allison 158, 222
Beam, George 124, 128, 182, 244
Bourke-White, Margaret 24
Brooks, Drex 106
Buckwalter, Harry H. 84, 96, 176, 199
Chase, Dana B. 34
Collier, John 88
Corsini, Harold 191
Davis, O. T. 126
Dawson, Robert H. 136, 166, 200, 212
E. K. Edwards 48
Farm Security Administration 17, 52, 102, 104, 158, 204, 222
Friedlander, Lee 202
Garrison, Ola Anfenson 77, 100
Gilpin, Laura 58, 74
Goin, Peter 54
Goodman, Charles 184

Grant 134
Gurnsey, B. H. 168
Harlan, Andrew James 50, 51
Helphand, Kenneth 20, 47, 108, 112, 170, 230, 248
Heyn 44
Hollard, L. 164
Jackson, William Henry xxii, 40, 41, 60, 61, 62, 70, 72, 86, 90, 178, 180, 198, 210, 216, 226, 227
Jennison, Andrea 36
Klett, Mark xxiii
Lee, Russell 102, 104, 204
Lillybridge, Charles S. 188
Manchester, Ellen 230
Matteson, S. W. 214
McClelland, Joe 224
McClure, L. C. 16, 22, 46, 76, 144, 146, 150, 152, 250
Mertin, Roger 208
Mollette, Rex 186

Myers, Joan 6
Nelson, Kathryn 256
Pagageorge, Tod 205
Photographer Unidentified 2, 18, 66, 94, 114, 118, 120, 132, 142, 154, 155, 160, 161, 190, 194, 220, 234, 238, 240, 258, 259, 260
Plowden, David 14, 42
Poley, Horace S. xiv
Regnier, L. D. 10, 172
Rephotographic Survey Project xxiii
Roach, O. 254
Rothstein, Arthur 52
Salisbury 252
Sturtevant, Joseph Bevier 116, 148, 196, 217
Tangen, Edward 32
Uelsmann, Jerry 56
Van Pelt, Richard 28
Verburg, JoAnn xxiii
Wolcott, Marion Post 17
Zellers, Robert G. 64, 98